HOLT

ChemFile
LAB PROGRAM

TEACHER'S
EDITION

Laboratory Experiments

A

HOLT, RINEHART AND WINSTON
Harcourt Brace & Company

Austin • New York • Orlando • Atlanta • San Francisco • Boston • Dallas • Toronto • London

Teacher's Safety Information **T5**

Master Materials List **T8**

    Chemicals **T8**

    Equipment **T11**

    Miscellaneous Materials **T13**

    Safety Equipment **T14**

Chemical Inventory **T15**

Incompatible Chemicals **T16**

Introduction .... **iv**

Safety .... **vi**

Labeling of Chemicals .... **xi**

Laboratory Techniques .... **xii**

A1   Laboratory Procedures **1**

A2   Accuracy and Precision in Measurements **21**

A3   Reactivity of Halide Ions **29**

A4   Test for Iron(II) and Iron(III) **33**

A5   Evidence for Chemical Change **37**

A6   Calcium and Its Compounds **43**

A7   Water of Hydration **49**

A8   Boyle's Law **55**

A9   Molar Volume of Gas **61**

A10   Molar Heat of Fusion of Ice **67**

A11   Ice-Nucleating Bacteria **71**

A12   Hydronium Ion Concentration and pH **77**

A13   Titration with an Acid and a Base **81**

A14   Energy and Entropy **87**

A15   Heat of Crystallization **93**

A16   Temperature of a Bunsen Burner Flame **97**

A17   Heat of Solution **103**

CONTENTS continued

A18 Heat of Combustion **109**

A19 The Solubility Product Constant
of Sodium Chloride **115**

A20 Buffering Capacity **119**

A21 Oxidation-Reduction Reactions **125**

A22 Cathodic Protection:
Factors Affecting the Corrosion of Iron **129**

A23 Carbon **135**

A24 Oil-Degrading Microbes **141**

A25 Polymers **147**

A26 Radioactivity **153**

A27 Detecting Radioactivity **157**

Laboratory Safety

HELPING STUDENTS RECOGNIZE THE IMPORTANCE OF SAFETY

One method that can help students appreciate the importance of precautions is to use a "safety contract" that students read and sign, indicating they have read, understand, and will respect the necessary safety procedures, as well as any other written or verbal instructions that will be given in class. You can find a copy of a safety contract on the *Holt ChemFile Teaching Resources CD-ROM*. You can use this form as a model or make your own safety contract for your students with language specific to your lab situation. When making your own contract you could include points such as the following.

- Make sure that students agree to always wear personal protective equipment (goggles and lab aprons). Safety information regarding the use of contact lenses is continually changing. Check your state and local regulations on this subject. Students should agree to read all lab exercises before they come to class. They should agree to follow all directions and safety precautions, and use only materials and equipment that you provide.
- Students should agree to remain alert and cautious at all times in the lab. They should never leave experiments unattended.
- Students should not wear heavy dangling jewelry or bulky clothing.
- Students should bring lab manuals and lab notebooks only into the lab. Backpacks, textbooks for other subjects, and other items elsewhere should be stored elsewhere.
- Students should agree to never eat, drink, or smoke in any science laboratory. Food should NEVER be brought into the laboratory
- Students should NEVER taste or touch chemicals.
- Students should keep themselves and other objects away from Bunsen-burner flames. Students should be responsible for checking that gas valves and hot plates are turned off before leaving the lab.
- Students should know the proper fire drill procedures and the locations of fire exits.
- Students should always clean all apparatus and work areas.
- Students should wash their hands thoroughly with soap and water before leaving the lab room.
- Students should know the location and operation of all safety equipment in the laboratory.
- Students should report all accidents or close calls to you immediately, no matter how minor.
- Students should NEVER work alone in the laboratory and they should never work unless you are present.

DISPOSAL OF CHEMICALS

Only a relatively small percentage of waste chemicals are classified as hazardous by EPA regulations. The EPA regulations are derived from two acts (as amended) passed by the Congress of the United States: RCRA (Resource Conservation and Recovery Act) and CERCLA (Comprehensive Environmental Response, Compensation, and Liability Act).

In addition, some states have enacted legislation governing the disposal of hazardous wastes that differs to some extent from the federal legislation. The disposal procedures described in this book have been designed to comply with the federal legislation as described in the EPA regulations.

In most cases the disposal procedures indicated in the teacher's edition will *probably* comply with your state's disposal requirements. However, to be sure of this, check with your state's environmental agency. If a particular disposal procedure does not comply with your state requirements, ask that office to assist you in devising a procedure that is in compliance.

The following general practices are recommended in addition to the specific instructions given in the margin notes of this Teacher's Edition.

- Except when otherwise specified in the disposal procedures, neutralize acidic and basic wastes with 1.0 M potassium hydroxide, KOH, or 1.0 M sulfuric acid, H_2SO_4, added slowly while stirring.
- In dealing with a waste-disposal contractor, prepare a complete list of the chemicals you want to dispose of. Classify each chemical on your disposal list as hazardous or nonhazardous waste. Check with your local environmental agency office for the details of such classification.
- Unlabeled bottles are a special problem. They must be identified to the extent that they can be classified as a hazardous or nonhazardous waste. Some landfills will analyze a mystery bottle for a fee if it is shipped to the landfill in a separate package, is labeled as a sample, and includes instructions to analyze the contents sufficiently to allow proper disposal.

ELECTRICAL SAFETY

Although none of the labs in this manual require electrical equipment, several include options for the use of microcomputer-based laboratory equipment, pH meters, or other equipment. The following safety precautions to avoid electric shocks must be observed any time electrical equipment is present in the lab.

- Each electrical socket in the laboratory must be a three-hole socket and must be protected with a GFI (ground-fault interrupter) circuit.
- Check the polarity of all circuits before use with a polarity tester from an electronics supply store. Repair any incorrectly wired sockets.
- Use only electrical equipment equipped with a three-wire cord and three-prong plug.
- Be sure all electrical equipment is turned off before it is plugged into a socket. Turn off electrical equipment before it is unplugged.
- Wiring hookups should be made or altered only when apparatus is disconnected from the power source and the power switch is turned off.
- Do not let electrical cords dangle from work stations; dangling cords are a shock hazard and students can trip over them.
- Do not use electrical equipment with frayed or twisted cords.
- The area under and around electrical equipment should be dry; cords should not lie in puddles of spilled liquid.
- Hands should be dry when using electrical equipment.

● Do not use electrical equipment powered by 110-115 V alternating current for conductivity demonstrations or for any other use in which bare wires are exposed, even if the current is connected to a lower voltage AC or DC connection.

Use dry cells or Ni-Cad rechargeable batteries as direct current sources. Do not use automobile storage batteries or AC-to-DC converters; these two sources of DC current can present serious shock hazards.

Prepared by Jay A. Young, Consultant, Chemical Health and Safety, Silver Spring, Maryland

Master Materials List

Chemical	Amount needed for 15 lab groups	Experiment
Acetic acid, glacial	12 mL	A12, A20
Acetone	75 mL	A25
Agar-agar (powdered)	30 g	A22
$AgNO_3$	8.5 g	A3
Alcohol, denatured	550 mL	A13, A22
$CaCl_2$	38 g	A6, A11
Calcium metal (small pieces)	45	A6
CaO	75 g	A6
$Ca(NO_3)_2$	41 g	A3
CH_3COONa	15.82 g	A19
$CH_3COONa \cdot 3H_2O$	13.6 g	A20
$Cu(NO_3)_2 \cdot 6H_2O$	320 g	A5, A21
$FeCl_3 \cdot 6H_2O$	41 g	A4, A21
$Fe(NH_4)_2(SO_4)_2 \cdot 6H_2O$	7.8 g	A4
Fertilizer	23 g	A24
$FeSO_4$	7.5 g	A21
HCl, concentrated	.784 L	A4, A12, A13, A20, A22, A23
HNO_3, concentrated	6.4 mL	A22
H_2SO_4, concentrated	56 mL	A21
H_3PO_4, concentrated	2 mL	A12
KBr	24 g	A3
KI	33 g	A3
$KMnO_4$	16 g	A21
KNO_3	10 g	A22
KSCN	1.9 g	A4
$K_3Fe(CN)_6$	34 g	A4, A22
$K_4Fe(CN)_6 \cdot 3H_2O$	21 g	A4
Magnesium sulfate, (Epsom salts) hydrated crystals	100 g	A7
$Mg(NO_3)_2 \cdot 2H_2O$	18 g	A21
NaCl	18 g	A1, A3, A12, A25
NaCl solution	150 mL	A19
NaF	5 g	A3

Chemical	Amount needed for 15 lab groups	Experiment
NaOCl, commercial bleach, 5%	5 mL	A3
NaOH	362 g	A5, A12, A13, A20, A22, A23, A27
$Na_2B_4O_7 \cdot 10H_2O$, solid borax	40 g	A25
$Na_2C_2O_4$	13 g	A22
Na_2CO_3	16 g	A12, A22
$Na_2S_2O_3 \cdot 5H_2O$	50 g	A3
$Na_3PO_4 \cdot 12H_2O$	38 g	A22
NH_3 (aq), concentrated	7 mL	A12
NH_4CH_3COO	8 g	A12
Penicillium culture	75 mL	A24
Phenolphthalein	11 g	A13, A22
Polyvinyl alcohol, solid	40 g	A25
Pseudomonas culture	75 mL	A24
$SnCl_2$, crystals	25 g	A21
Sodium polyacrylate	15 salt shakers	A25
Sodium thiosulfate pentahydrate, seed crystals	75 g	A14
Soluble starch	3 g	A3
$Zn(NO_3)_2$	30 g	A21

Equipment

Equipment	Amount needed for 15 lab groups	Experiment
Balance	15	A1, A2, A7, A14, A15, A16, A19, A21, A23
Beaker, 50 mL	30	A1, A6, A9, A20, A21, A25
Beaker, 100 mL	45	A2, A5, A13
Beaker, 150 mL	30	A4, A19
Beaker, 250 mL	30	A1, A6, A12, A23
Beaker, 400 mL	30	A9, A10, A15, A20, A22
Beaker, 600 mL	30	A14
Beaker tongs	15	A1
Boyle's law apparatus	15	A8
Bunsen burner with gas tubing and sparker (a hot plate may be substituted in some cases)	15	A1, A5, A6, A7, A10, A14, A15, A16, A19, A22, A23
Buret clamp	15	A1, A9, A13, A14, A23
Burets, 50 mL capacity	30	A1, A9, A13, A23
Celsius thermometer, nonmercury, range −10 to 120 °C	30	A1, A2, A9, A10, A14, A15, A17, A16
Crucible tongs	15	A10, A18
Crucible and lid (metal or ceramic)	15	A7, A23
Density indicator strips	45	A24
Desiccator	15	A7
Evaporating dish	15	A1, A19
Eudiometer, 50 mL	15	A9
Flask, Erlenmeyer 125 mL	30	A12, A23
Forceps	15	A1, A14, A15, A21, A23
Funnels, glass	15	A1, A23
Funnel rack	15	A6, A23
Glass stirring rod	15	A5, A10, A12, A15, A19, A22, A25

Equipment	Amount needed for 15 lab groups	Experiment
Glass tubing		A6
Graduated cylinder, 10 mL	15	A6, A12, A19, A20, A25
Graduated cylinder, 50 mL	15	A6, A12, A19, A23
Graduated cylinder, 100 mL	15	A1, A10, A15, A18
Graduated cylinder, 1000 mL	1	A9
Medicine dropper	15	A6, A21, A25
Metric ruler	15	A2, A6, A27
Petri dish with lid	30	A22, A25
pH meter or CBL with pH probe	15	A20
pH paper, wide and narrow range	1 roll each	A12, A20
Pipe-stem triangle	15	A7
Radioactivity scaler	15	A26
Ring , iron	15	A1, A5, A6, A7, A10, A14, A15, A16, A18, A19, A23
Ring stand	15	A1, A5, A6, A7, A10, A13, A14, A15, A16, A18, A19, A22, A23
Rubber stopper (for test tubes)	75	A6, A23
Rubber stopper (one-hole)	15	A9
Spatula	15	A6, A7
Test tube (small)	180	A3, A5, A12, A22, A23
Test tube (large)	15	A6, A14, A15, A19, A21, A23
Test-tube clamp	15	A6, A15, A23
Test-tube rack	15	A6, A14, A22
Thermometer clamp	15	A18
Tongs	15	A7, A19, A22
Wash bottle	15	A1, A13
Watch glass	15	A25

Equipment	Amount needed for 15 lab groups	Experiment
Weighing paper	15	A1
Wire gauze (nonabestos) with ceramic center	15	A1, A5, A10, A14, A15, A19, A22
Wire stirrer	15	A14

Miscellaneous Materials

Material	Amount needed for 15 lab groups	Experiment
Aluminum foil	1 roll	A26
Candle	15	A18
Clock with second hand or other timing device	1	A14
Etch clamp (ring from a keychain)	15	A27
Filter paper	230	A1, A6
Heat-resistant mat	15	A1
Ice	cubes	A10, A18
Index cards	2 packs	A26, A27
Liquid soap	bottle	A6
Markers, fine-tipped permanent	15	A5
Matches	box	A6
Metal objects (copper, iron, or nickel) 20–30 g samples of each	15	A16
Molecular model kit	30	A23
Nichrome wire	300 cm	A16
Plastic, CR-39	15	A27
Plastic cups (foam)	30	A10, A15, A16
Plastic cups (transparent)	15	A27
Pliers	15	A22
Push pin	15	A27
Sand paper	15 sheets	A21
Scissors	15	A27
Tape (transparent)	1 roll	A27
Tin can, 10 oz.	15	A18
Tin can, 46 oz.	15	A18
Tin can lid	15	A18
Tissue paper	1 roll	A26
Wire, aluminum	180 cm	A5
Wire, copper, 18 gauge	150 cm	A1
Wooden splints	15	A6

Safety Equipment

Equipment	Amount needed for 15 lab groups	Experiment
Face shield	1	Teacher use only
Impermeable gloves	1 pair	Teacher use only
Lab apron	30	all
Safety goggles	30 pair	all

Chemical Inventory

There are many computer programs and data bases that you can use to track your chemical inventory. The following is an example of the information you could include within your inventory tracking system.

School Name: _____

Street Address: _____

City/County/Zip Code: _____

Chemical name and concentration or form	Estimated amount on hand	Amount and date purchased	Comments
Acetic acid, 5% solution	250 mL	2 L 9/97	Can substitute white vinegar
Acetic acid, glacial	900 mL	2 L 10/96	
$AgNO_3$, solid, technical grade	300 g	500 g 4/95	
$AgNO_3$, 1 M solution	200 mL	500 mL 9/97	Store in amber bottles
$Ba(NO_3)_2$, solid, technical grade	40 g	100 g 4/95	
$BaCl_2$, anhydrous form	130 g	250 g 4/95	
$BaCl_2 \cdot 5H_2O$, solid	15 g	100 g 4/95	Can substitute the anhydrous form

Incompatible Chemicals

The following listing should be considered when organizing and storing chemicals. Note that some chemicals on this list should no longer be in your lab due to their potential risks. Consult your state and local guidelines for more specific information on chemical hazard:

Chemical	Should not come in contact with
Acetic acid	Chromic acid, nitric acid, perchloric acid, ethylene glycol, hydroxyl compounds, peroxides, and permanganates
Acetone	Concentrated sulfuric acid and nitric acid mixtures
Acetylene	Silver, mercury and their compounds; bromine, chlorine, fluorine, and copper tubing
Alkali metals, powdered aluminum and magnesium	Water, carbon dioxide, carbon tetrachloride, and the halogens

Chemical	Should not come in contact with
Ammonia (anhydrous)	Mercury, hydrogen fluoride, and calcium hypochlorite
Ammonium nitrate (strong oxidizer)	Strong acids, metal powders, chlorates, sulfur, flammable liquids, and finely-divided organic materials
Aniline	Nitric acid and hydrogen peroxide
Bromine	Ammonia, acetylene, butane, hydrogen, sodium carbide, turpentine, and finely-divided metals
Carbon (activated)	Calcium hypochlorite, all oxidizing agents
Chlorates	Ammonium salts, strong acids, powdered metals, sulfur, and finely-divided organic materials
Chromic acid	Glacial acetic acid, camphor, glycerin, naphthalene, turpentine, low-molar mass alcohols, and flammable liquids
Chlorine	Same as bromine
Copper	Acetylene and hydrogen peroxide
Flammable liquids	Ammonium nitrate, chromic acid, hydrogen peroxide, sodium peroxide, nitric acid, and the halogens
Hydrocarbons (butane, propane, gasoline, turpentine)	Fluorine, chlorine, bromine, chromic acid, and sodium peroxide
Hydrofluoric acid	Ammonia
Hydrogen peroxide	Most metals and their salts, flammable liquids, and other combustible materials
Hydrogen sulfide	Nitric acid and certain other oxidizing gases
Iodine	Acetylene and ammonia
Nitric acid	Glacial acetic acid, chromic and hydrocyanic acids, hydrogen sulfide, flammable liquids, and flammable gases that are easily nitrated
Oxygen	Oils, grease, hydrogen, flammable substances
Perchloric acid	Acetic anhydride, bismuth and its alloys, alcohols, paper, wood, and other organic materials
Phosphorus pentoxide	Water
Potassium permanganate	Glycerin, ethylene glycol, and sulfuric acid
Silver	Acetylene, ammonium compounds, oxalic acid, and tartaric acid
Sodium peroxide	Glacial acetic acid, acetic anhydride, methanol, carbon disulfide, glycerin, benzaldehyde, and water
Sulfuric acid	Chlorates, perchlorates, permanganates, and water

ChemFile LAB A

Laboratory
Experiments

HOLT, RINEHART AND WINSTON
Harcourt Brace & Company
Austin • New York • Orlando • Atlanta • San Francisco • Boston • Dallas • Toronto • London

Cover: HRW Photo by Sam Dudgeon

Printed in the United States of America

ISBN 0-03-051928-4

9 10 11 12 13 14 15 022 05 04 03 02 01

LAB MANUAL **A** CONTENTS

Introduction v

Safety vii

Labeling of Chemicals xi

Laboratory Techniques xii

A1 Laboratory Procedures 1

A2 Accuracy and Precision in Measurements 21

A3 Reactivity of Halide Ions 29

A4 Test for Iron(II) and Iron(III) 33

A5 Evidence for Chemical Change 37

A6 Calcium and Its Compounds 43

A7 Water of Hydration 49

A8 Boyle's Law 55

A9 Molar Volume of Gas 61

A10 Molar Heat of Fusion of Ice 67

A11 Ice-Nucleating Bacteria 71

A12 Hydronium Ion Concentration and pH 77

A13 Titration with an Acid and a Base 81

A14 Energy and Entropy 87

A15 Heat of Crystallization 93

A16 Temperature of a Bunsen Burner Flame 97

A17 Heat of Solution 103

A18 Heat of Combustion 109

A19 The Solubility Product Constant
of Sodium Chloride 115

A20 Buffering Capacity 119

A21 Oxidation-Reduction Reactions 125

A22 Cathodic Protection:
Factors Affecting the Corrosion of Iron 129

A23 Carbon 135

A24 Oil-Degrading Microbes 141

CONTENTS continued

A25 Polymers 147

A26 Radioactivity 153

A27 Detecting Radioactivity 157

Introduction to the Lab Program

STRUCTURE OF THE EXPERIMENTS

INTRODUCTION

The opening paragraphs set the theme for the experiment and summarize its major concepts.

OBJECTIVES

Objectives highlight the key concepts to be learned in the experiment and emphasize the science process skills and techniques of scientific inquiry.

MATERIALS

These lists enable you to organize all apparatus and materials needed to perform the experiment. Knowing the concentrations of solutions is vital. You often need this information to perform calculations and to answer the questions at the end of the experiment.

SAFETY

Safety cautions are placed at the beginning of the experiment to alert you to procedures that may require special care. Before you begin, you should review with the safety issues that apply to the experiment.

PROCEDURE

By following the procedures of an experiment, you are performing concrete laboratory operations that duplicate the fact-gathering techniques used by professional chemists. You are learning skills in the laboratory. The procedures tell you how and where to record observations and data.

DATA AND CALCULATIONS TABLES

The data that you will collect during each experiment should be recorded in the labeled Data Tables provided. The entries you make in a Calculations Table emphasize the mathematical, physical, and chemical relationships that exist among the accumulated data. Both tables should help them to think logically and to formulate their conclusions about what occurred during the experiment.

CALCULATIONS

Space is provided for all computations based on the data you have gathered.

QUESTIONS

Based on the data and calculations, you should be able to develop a plausible explanation for the phenomena observed during the experiment. Specific questions are asked that require you to draw on the concepts you have learned.

GENERAL CONCLUSIONS

This section asks broader questions that bring together the results and conclusions of the experiment and relate them to other situations.

Safety in the Chemistry Laboratory

Chemicals are not toys

Any chemical can be dangerous if it is misused. Always follow the instructions for the experiment. Pay close attention to the safety notes. Do not do anything differently unless told to do so by your teacher.

Chemicals, even water, can cause harm. The trick is to know how to use chemicals correctly so that they will not cause harm. If you follow the rules stated in the following pages, pay attention to your teacher's directions, and follow the cautions on chemical labels and the experiments, then you will be using chemicals correctly.

These safety rules always apply in the lab

1. **Always wear a lab apron and safety goggles.**
   Even if you aren't working on an experiment, laboratories contain chemicals that can damage your clothing, so wear your apron and keep the strings of the apron tied. Because chemicals can cause eye damage, even blindness, you must wear safety goggles. If your safety goggles are uncomfortable or get clouded up, ask your teacher for help. Try lengthening the strap a bit, washing the goggles with soap and warm water, or using an antifog spray.

2. **No contact lenses are allowed in the lab.**
   Even while wearing safety goggles, chemicals could get between contact lenses and your eyes and cause irreparable eye damage. If your doctor requires that you wear contact lenses instead of glasses, then you should wear eye-cup safety goggles in the lab. Ask your doctor or your teacher how to use this very important and special eye protection.

3. **Never work alone in the laboratory.**
   You should always do lab work only under the supervision of your teacher.

4. **Wear the right clothing for lab work.**
   Necklaces, neckties, dangling jewelry, long hair, and loose clothing can cause you to knock things over or catch items on fire. Tuck in neckties or take them off. Do not wear a necklace or other dangling jewelry, including hanging earrings. It isn't necessary, but it might be a good idea to remove your wristwatch so that it is not damaged by a chemical splash.

   Pull back long hair, and tie it in place. Nylon and polyester fabrics burn and melt more readily than cotton, so wear cotton clothing if you can. It's best to wear fitted garments, but if your clothing is loose or baggy, tuck it in or tie it back so that it does not get in the way or catch on fire.

   Wear shoes that will protect your feet from chemical spills—no open-toed shoes or sandals and no shoes with woven leather straps. Shoes made of solid leather or a polymer are much better than shoes made of cloth. Also, wear pants, not shorts or skirts.

5. **Only books and notebooks needed for the experiment should be in the lab.**
   Do not bring other textbooks, purses, bookbags, backpacks, or other items into the lab; keep these things in your desk or locker.

6. **Read the entire experiment before entering the lab.**
   Memorize the safety precautions. Be familiar with the instructions for the experiment. Only materials and equipment authorized by your teacher should be used. When you do the lab work, follow the instructions and the safety precautions described in the directions for the experiment.

7. **Read chemical labels.**
   Follow the instructions and safety precautions stated on the labels. Know the location of Materials Safety Data Sheets for chemicals

8. **Walk carefully in the lab.**
   Sometimes you will carry chemicals from the supply station to your lab station. Avoid bumping other students and spilling the chemicals. Stay at your lab station at other times.

9. **Food, beverages, chewing gum, cosmetics, and smoking are NEVER allowed in the lab.**
   You already know this.

10. **Never taste chemicals or touch them with your bare hands.**
    Also, keep your hands away from your face and mouth while working, even if you are wearing gloves.

11. **Use a sparker to light a Bunsen burner.**
    Do not use matches. Be sure that all gas valves are turned off and that all hot plates are turned off and unplugged when you leave the lab.

12. **Be careful with hot plates, Bunsen burners, and other heat sources.**
    Keep your body and clothing away from flames. Do not touch a hot plate after it has just been turned off. It is probably hotter than you think. The same is true of glassware, crucibles, and other things after you remove them from a hot plate, drying oven, or the flame of a Bunsen burner.

13. **Do not use electrical equipment with frayed or twisted cords or wires.**

14. **Be sure your hands are dry before using electrical equipment**
    Before plugging an electrical cord into a socket, be sure the electrical equipment is turned off. When you are finished with it, turn it off. Before you leave the lab, unplug it, but be sure to turn it off first.

15. **Do not let electrical cords dangle from work stations; dangling cords can cause tripping or electrical shocks.**
    The area under and around electrical equipment should be dry; cords should not lie in puddles of spilled liquid.

16. **Know fire drill procedures and the locations of exits.**

17. **Know the location and operation of safety showers and eyewash stations.**

18. **If your clothes catch on fire, walk to the safety shower, stand under it, and turn it on.**

19. **If you get a chemical in your eyes, walk immediately to the eyewash station, turn it on, and lower your head so that your eyes are in the running water.**
    Hold your eyelids open with your thumbs and fingers, and roll your eyeballs around. You have to flush your eyes continuously for at least 15 mm. Call your teacher while you are doing this.

20. **If you have a spill on the floor or lab bench, call your teacher rather than trying to clean it up by yourself.**
Your teacher will tell you if it is OK for you to do the cleanup; if it is not, your teacher will know how the spill should be cleaned up safely.

21. **If you spill a chemical on your skin, wash it off under the sink faucet, and call your teacher.**
If you spill a solid chemical on your clothing, brush it off carefully so that you do not scatter it, and call your teacher. If you get a liquid on your clothing, wash it off right away if you can get it under the sink faucet, and call your teacher. If the spill is on clothing that will not fit under the sink faucet, use the safety shower. Remove the affected clothing while under the shower, and call your teacher. (It may be temporarily embarrassing to remove your clothing in front of your class, but failing to flush that chemical off your skin could cause permanent damage.)

22. **The best way to prevent an accident is to stop it before it happens.**
If you have a close call, tell your teacher so that you and your teacher can find a way to prevent it from happening again. Otherwise, the next time, it could be a harmful accident instead of just a close call.

23. **All accidents should be reported to your teacher, no matter how minor.**
Also, if you get a headache, feel sick to your stomach, or feel dizzy, tell your teacher immediately.

24. **For all chemicals, take only what you need.**
On the other hand, if you do happen to take too much and have some left over, DO NOT put it back in the bottle. If somebody accidentally puts a chemical into the wrong bottle, the next person to use it will have a contaminated sample. Ask your teacher what to do with any leftover chemicals.

25. **NEVER take any chemicals out of the lab.**
You should already know this rule.

26. **Horseplay and fooling around in the lab are very dangerous.**
NEVER be a clown in the laboratory.

27. **Keep your work area clean and tidy.**
After your work is done, clean your work area and all equipment.

28. **Always wash your hands with soap and water before you leave the lab.**

29. **Whether or not the lab instructions remind you, ALL of these rules APPLY ALL OF THE TIME.**

QUIZ
Determine which safety rules apply to the following.

- Tie back long hair, and confine loose clothing. (Rule ? applies.)
- Never reach across an open flame. (Rule ? applies.)
- Use proper procedures when lighting Bunsen burners. Turn off hot plates, Bunsen burners, and other heat sources when not in use. (Rule ? applies.)
- Heat flasks and beakers on a ring stand with wire gauze between the glass and the flame. (Rule ? applies.)
- Use tongs when heating containers. Never hold or touch containers with your hands while heating them. Always allow heated materials to cool before handling them. (Rule ? applies.)
- Turn off gas valves when not in use. (Rule ? applies.)

ChemFile LAB A

SAFETY SYMBOLS

To highlight specific types of precautions, the following symbols are used in the experiments. Remember that no matter what safety symbols and instructions appear in each experiment, all of the 29 safety rules described previously should be followed at all times.

Eye and clothing protection

- Wear laboratory aprons in the laboratory. Keep the apron strings tied so that they do not dangle.
- Wear safety goggles in the laboratory at all times. Know how to use the eyewash station.

Chemical safety

- Never taste, eat, or swallow any chemicals in the laboratory. Do not eat or drink any food from laboratory containers. Beakers are not cups, and evaporating dishes are not bowls.
- Never return unused chemicals to the original container.
- Some chemicals are harmful to the environment. You can help protect the environment by following the instructions for proper disposal.
- It helps to label the beakers and test tubes containing chemicals.
- Never transfer substances by sucking on a pipette or straw; use a suction bulb.
- Never place glassware, containers of chemicals, or anything else near the edges of a lab bench or table.

Caustic substances

- If a chemical gets on your skin or clothing or in your eyes, rinse it immediately, and alert your teacher.
- If a chemical is spilled on the floor or lab bench, tell your teacher, but do not clean it up yourself unless your teacher says it is OK to do so.

Heating safety

- When heating a chemical in a test tube, always point the open end of the test tube away from yourself and other people. (This is another new rule.)

Explosion precaution

- Use flammable liquids only in small amounts.
- When working with flammable liquids, be sure that no one else in the lab is using a lit Bunsen burner or plans to use one. Make sure there are no other heat sources present.

Hand safety

- Always wear gloves or cloths to protect your hands when cutting, fire polishing, or bending hot glass tubing. Keep cloths clear of any flames.
- Never force glass tubing into rubber tubing, rubber stoppers, or corks. To protect your hands, wear heavy leather gloves or wrap toweling around the glass and the tubing, stopper, or cork, and gently push in the glass tubing.
- Use tongs when heating test tubes. Never hold a test tube in your hand to heat it.
- Always allow hot glasswear to cool before handling.

Glassware safety

- Check the condition of glassware before and after using it. Inform your teacher of any broken, chipped, or cracked glassware because it should not be used.
- Do not pick up broken glass with your bare hands. Place broken glass in a specially designated disposal container.

Gas precaution

- Do not inhale fumes directly. When instructed to smell a substance, use your hand, wave the fumes toward your nose, and inhale gently. (Some people say "waft the fumes.")

Radiation precaution

- Always wear gloves when handling a radioactive source.
- Always wear safety goggles when performing experiments with radioactive materials.
- Always wash your hands and arms thoroughly after working with radioactive materials.

Hygienic care

- Keep your hands away from your face and mouth.
- Always wash your hands before leaving the laboratory.

Any time you see any of the safety symbols you should remember that all 29 of the numbered laboratory rules always apply.

Labeling of Chemicals

In any science laboratory the *labeling* of chemical containers, reagent bottles, and equipment is essential for safe operations. Proper labeling can lower the potential for accidents that occur as a result of misuse. Labels and equipment instructions should be read several times before using. Be sure that you are using the correct items, that you know how to use them, and that you are aware of any hazards or precautions associated with their use.

All chemical containers and reagent bottles should be labeled prominently and accurately using labeling materials that are not affected chemicals.

Chemical labels should contain the following information.

1. **Name of chemical and the chemical formula**
2. **Statement of possible hazards** This is indicated by the use of an appropriate signal word, such as DANGER, WARNING, or CAUTION. This signal word usually is accompanied by a word that indicates the type of hazard present such as POISON, CAUSES BURNS, EXPLOSIVE or FLAMMABLE. Note that this labeling should not take the place of reading the appropriate Material Safety Data Sheet for a chemical.
3. **Precautionary measures** Precautionary measures describe how users can avoid injury from the hazards listed on the label. Examples include: "Use only with adequate ventilation," and "Do not get in eyes or on skin or clothing."
4. **Instructions in case of contact or exposure** If accidental contact or exposure does occur immediate treatment is often necessary to minimize injury. Such treatment usually consists of proper first-aid measures that can be used before a physician administers treatment. An example is: "In case of contact, flush with large amounts of water; for eyes, rinse freely with water for 15 minutes and get medical attention immediately"
5. **The date of preparation and the name of the person who prepared the chemical** This information is important for maintaining a safe chemical inventory.

Suggested Labeling Scheme

Name of contents	Hydrochloric Acid	
	6 M HCl	Chemical formula and concentration or physical state
Statements of possible hazards and precautionary measures	WARNING! CAUSTIC and CORROSIVE-CAUSES BURNS CAUTION! Avoid contact with skin and eyes. Avoid breathing vapors.	
	IN CASE OF CONTACT: Immediately flush skin or eyes with large amounts of water for at least 15 minutes; for eyes, get medical attention immediately!	Hazard Instructions for contact or overexposure
Date prepared or obtained	May 8, 1989 Prepared by Betsy Byron Faribault High School, Faribault, Minnesota	Manufacturer (Commercially obtained) or preparer (Locally made)

Laboratory Techniques

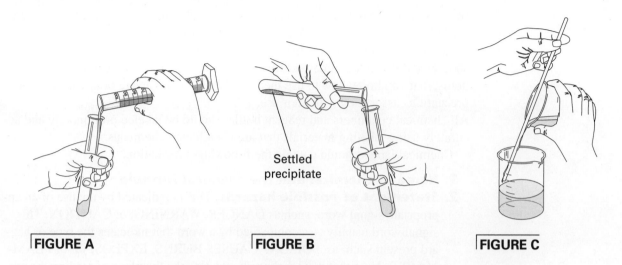

Settled precipitate

FIGURE A   FIGURE B   FIGURE C

DECANTING AND TRANSFERRING LIQUIDS

1. The safest way to transfer a liquid from a graduated cylinder to a test tube is shown in Figure A. The liquid is transferred at arm's length with the elbows slightly bent. This position enables you to see what you are doing and still maintain steady control.

2. Sometimes liquids contain particles of insoluble solids that sink to the bottom of a test tube or beaker. Use one of the methods shown below to separate a supernatant (the clear fluid) from insoluble solids.

a. Figure B shows the proper method of decanting a supernatant liquid in a test tube.

b. Figure C shows the proper method of decanting a supernatant liquid in a beaker by using a stirring rod. The rod should touch the wall of the receiving container. Hold the stirring rod against the lip of the beaker containing the supernatant liquid. As you pour, the liquid will run down the rod and fall into the beaker resting below. Using this method, the liquid will not run down the side of the beaker from which you are pouring.

HEATING SUBSTANCES AND EVAPORATING SOLUTIONS

1. Use care in selecting glassware for high-temperature heating. The glassware should be heat resistant.

2. When heating glassware using a gas flame, use a ceramic-centered wire gauze to protect glassware from direct contact with the flame. Wire gauzes can withstand extremely high temperatures and will help prevent glassware from breaking. Figure D shows the proper setup for evaporating a solution over a water bath.

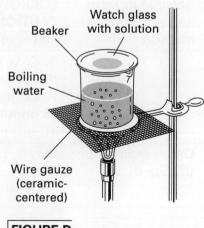

Watch glass with solution

Beaker

Boiling water

Wire gauze (ceramic-centered)

FIGURE D

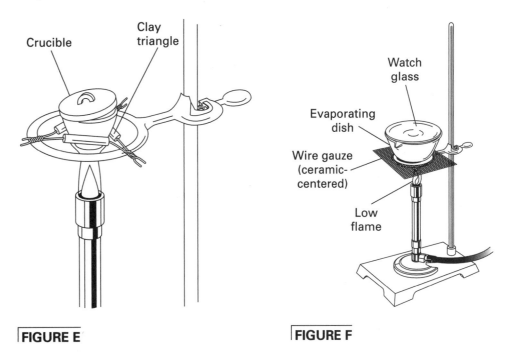

Crucible

Clay triangle

Watch glass

Evaporating dish

Wire gauze (ceramic-centered)

Low flame

FIGURE E

FIGURE F

3. In some experiments you are required to heat a substance to high temperatures in a porcelain crucible. Figure E shows the proper apparatus setup used to accomplish this task.

4. Figure F shows the proper setup for evaporating a solution in a porcelain evaporating dish with a watch glass cover that prevents spattering.

5. Glassware, porcelain, and iron rings that have been heated may *look* cool after they are removed from a heat source, but these items can still burn your skin even after several minutes of cooling. Use tongs, test-tube holders, or heat-resistant mitts and pads whenever you handle this apparatus.

6. You can test the temperature of questionable beakers, ring stands, wire gauzes, or other pieces of apparatus that have been heated by holding the back of your hand close to their surfaces before grasping them. You will be able to feel any heat generated from the hot surfaces. DO NOT TOUCH THE APPARATUS. Allow plenty of time for the apparatus to cool before handling.

HOW TO POUR LIQUID FROM A REAGENT BOTTLE

1. Read the label at least three times before using the contents of a reagent bottle.

2. Never lay the stopper of a reagent bottle on the lab table.

3. When pouring a caustic or corrosive liquid into a beaker use a stirring rod to avoid drips and spills. Hold the stirring rod against the lip of the reagent bottle. Estimate the amount of liquid you need and pour this amount along the rod into the beaker. See Figure G.

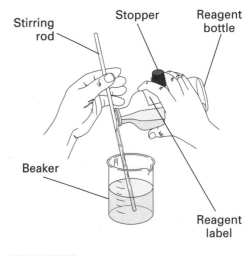

Stirring rod

Stopper

Reagent bottle

Beaker

Reagent label

FIGURE G

4. Extra precaution should be taken when handling a bottle of acid. Remember the following important rules: Never add water to any concentrated acid, particularly sulfuric acid, because the mixture can splash and will generate a lot of heat. To dilute any acid, add the acid to water in small quantities, while stirring slowly. Remember the "triple A's"-Always Add Acid to water.

5. Examine the outside of the reagent bottle for any liquid that has dripped down the bottle or spilled on the counter top. Your teacher will show you the proper procedures for cleaning up a chemical spill.

6. Never pour reagents back into stock bottles. At the end of the experiment, your teacher will tell you how to dispose of any excess chemicals.

HOW TO HEAT MATERIAL IN A TEST TUBE

1. Check to see that the test tube is heat-resistant.

2. Always use a test tube holder or clamp when heating a test tube.

3. Never point a heated test tube at anyone, because the liquid may splash out of the test tube.

4. Never look down into the test tube while heating it.

5. Heat the test tube from the upper portions of the tube downward and continuously move the test tube as shown in Figure H. Do not heat any one spot on the test tube. Otherwise a pressure build-up may cause the bottom of the tube to blow out.

HOW TO USE A MORTAR AND PESTLE

1. A mortar and pestle should be used for grinding only one substance at a time. See Figure I.

2. Never use a mortar and pestle for simultaneously mixing different substances.

3. Place the substance to be broken up into the mortar

4. Pound the substance with the pestle and grind to pulverize.

5. Remove the powdered substance with a porcelain spoon.

DETECTING ODORS SAFELY

1. Test for the odor of gases by wafting your hand over the test tube and cautiously sniffing the fumes as shown in Figure J.

2. Do not inhale any fumes directly.

3. Use a fume hood whenever poisonous or irritating fumes are evolved. DO NOT waft and sniff poisonous or irritating fumes.

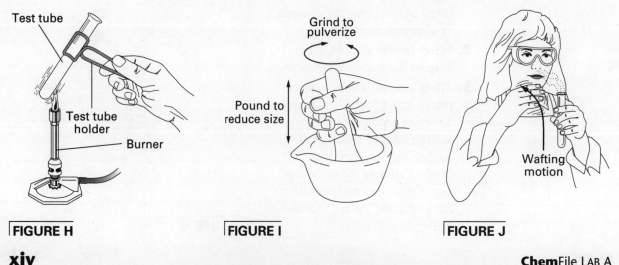

FIGURE H | **FIGURE I** | **FIGURE J**

ChemFile LAB A

Name _____

Date _____ Class _____

HOLT
ChemFile
LAB PROGRAM

EXPERIMENT **A1**

Laboratory Procedures

OBJECTIVES

Recommended time:
2 lab periods

- **Observe** proper safety techniques with all laboratory equipment.
- **Use** laboratory apparatus skillfully and efficiently.
- **Recognize** the names and functions of all apparatus in the laboratory.
- **Develop** a positive approach toward laboratory safety.

INTRODUCTION

The best way to become familiar with chemical apparatus is to handle the pieces yourself in the laboratory. This experiment is divided into several parts in which you will learn how to adjust the gas burner, insert glass tubing into a rubber stopper, use a balance, handle solids, measure liquids, filter a mixture, and measure temperature and heat. Great emphasis is placed on safety precautions that should be observed whenever you perform an experiment and use the apparatus. Several useful manipulative techniques are also illustrated on pages xii through xiv. In many of the later experiments, references will be made to these "Laboratory Techniques." In later experiments you will also be referred to the safety precautions and procedures explained in all parts of this experiment. It is important that you develop a positive approach to a safe and healthful environment in the lab.

SAFETY

Required Precautions

- Discuss all safety symbols and caution statements with students.

- Safety goggles and a lab apron must be worn at all times.

- Read all safety cautions and discuss them with your students.

- Loose hair and clothing must be tied back.

- If students find a gas valve that has not been turned off properly, the room must be thoroughly ventilated, using the hood and windows in order to prevent a fire hazard.

Always wear safety goggles and a lab apron to protect your eyes and clothing. If you get a chemical in your eyes, immediately flush the chemical out at the eyewash station while calling to your teacher. Know the location of the emergency lab shower and eyewash station and the procedure for using them.

Do not touch any chemicals used in the laboratory. If you get a chemical on your skin or clothing, wash the chemical off at the sink while calling to your teacher. Make sure you carefully read the labels and follow the precautions on all containers of chemicals that you use. If there are no precautions stated on the label, ask your teacher what precautions to follow. Never return leftover chemicals to their original containers; take only small amounts to avoid wasting supplies.

Never put broken glass into a regular waste container. Broken glass should be disposed of separately according to your teacher's instructions.

HRW material copyrighted under notice appearing earlier in this work.

When using a Bunsen burner, confine long hair and loose clothing. If your clothing catches on fire, WALK to the emergency lab shower, and use it to put out the fire. Do not heat glassware that is broken, chipped, or cracked. Use tongs or a hot mitt to handle heated glassware and other equipment because heated glassware does not look hot.

 When you insert glass tubing into stoppers, lubricate the glass with water or glycerin and protect your hands and fingers: Wear leather gloves or place folded cloth pads between both of your hands and the glass tubing. Then *gently* push the tubing into the stopper hole. In the same way, protect your hands and fingers when removing glass tubing from stoppers and from rubber or plastic tubing.

MATERIALS

PART 1   THE BURNER

- Bunsen burner and related equipment
- copper wire, 18 gauge
- evaporating dish
- forceps
- heat-resistant mat
- sparker

PROCEDURE

Procedural Tips

• Although all the techniques are carefully described, there is no substitute for seeing each of them demonstrated before students try them themselves.

1. The Bunsen burner is commonly used as a source of heat in the laboratory. Look at Figure A as you examine your Bunsen burner and identify the parts. Although the details of construction vary among burners, each has a gas inlet located in the base, a vertical tube or barrel in which the gas is mixed with air, and adjustable openings or ports in the base of the barrel. These ports admit air to the gas stream. The burner may have an adjustable needle valve to regulate the flow of gas. In some models the gas flow is regulated simply by adjusting the gas valve on the supply line. The burner is always turned off at the gas valve, never at the needle valve.

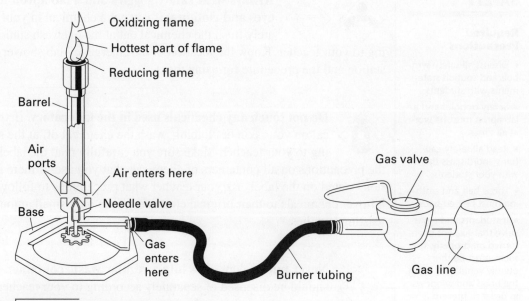

FIGURE A

• Close supervision is necessary if students work with glass tubing. Emphasize that they should never attempt to force tubing into stoppers with excessive force. Advise them to use a lubricant such as water or glycerin to help ease the tubing in.

• Stress to students that hot tubing or other glassware does not look hot, and care should be taken to avoid burns.

• Advise students to read and follow the procedures carefully.

Disposal

• There are no special disposal guidelines. The sodium chloride can be reused by each class if you designate a special container for NaCl disposal after each part. Keep this NaCl separate from the NaCl you use as a reagent. The same is true for the copper wire and the sand.

CAUTION Before you light the burner, check to see that you and your partner have taken the following safety precautions against fires: Wear safety goggles, aprons, and gloves. Confine long hair and loose clothing: tie long hair at the back of the head and away from the front of the face, and roll up long sleeves on shirts, blouses, and sweaters away from the wrists. You should also know the locations of fire extinguishers, fire blankets, safety showers, and sand buckets and the procedure for using them in case of a fire.

2. When lighting the burner, partially close the ports at the base of the barrel, turn the gas full on, hold the sparker about 5 cm above the top of the burner, and proceed to light. The gas flow may then be regulated by adjusting the gas valve until the flame has the desired height. If a very low flame is needed, remember that the ports should be partially closed when the gas pressure is reduced. Otherwise the flame may burn inside the base of the barrel. When the flame is improperly burning in this way, the barrel will get very hot, and the flame will produce a poisonous gas, carbon monoxide.

CAUTION If the flame is burning inside the base of the barrel, immediately turn off the gas at the gas valve. Do not touch the barrel, for it is extremely hot. Allow the barrel of the burner to cool and then proceed as follows:

Begin again, but first decrease the amount of air admitted to the burner by partially closing the ports. Turn the gas full on and then relight the burner. Control the height of the flame by adjusting the gas valve. By taking these steps, you should acquire a flame that is burning safely and is easily regulated.

3. Once you have a flame that is burning safely and steadily, you can experiment by completely closing the ports at the base of the burner. What effect does this have on the flame?

When the ports at the base are closed, a yellow flame results.

Using the forceps, hold an evaporating dish in the tip of the flame for about 3 min. Place the dish on a heat-resistant mat and allow the dish to cool. Then examine the bottom of the dish. Describe the results and suggest a possible explanation.

The yellow flame leaves a black deposit on the bottom of the

evaporating dish. The lack of air to the burner results in incom-

plete combustion, the luminous flame, and the deposit of soot.

Such a flame is seldom used in the lab. For laboratory work, you should adjust the burner so that the flame is free of yellow color, nonluminous, and also free of the roaring sound caused by admitting too much air.

4. Regulate the flow of gas so that the flame extends roughly 8 cm above the barrel. Now adjust the supply of air until you have a quiet, steady flame with

a sharply defined, light blue inner cone. This adjustment gives the highest temperature possible with your burner. Using the forceps, insert a 10 cm piece of copper wire into the flame just above the barrel. Lift the wire slowly up through the flame. Where is the hottest portion of the flame located?

The hottest portion of the nonluminous flame is just above the

tip of the light blue cone.

Hold the wire in this part of the flame for a few seconds. What happens?

The copper wire becomes red hot and begins to soften (melt).

5. Shut off the gas burner. Now think about what you have just observed in steps **3** and **4.** Why is the nonluminous flame preferred over the yellow luminous flame in the laboratory?

The nonluminous flame burns cleaner than the yellow flame.

6. Clean the evaporating dish and put away the burner. All the equipment you store in the lab locker or drawer should be completely cool, clean, and dry. Be sure that the valve on the gas jet is completely shut off. Remember to wash your hands thoroughly with soap at the end of each laboratory period.

MATERIALS

PART 2 GLASS MANIPULATIONS

- cloth pads or leather gloves
- glass funnel
- rubber hose
- rubber stopper, 1-hole
- water or glycerin

PROCEDURE

1. Inserting glass tubing into rubber stoppers can be very dangerous. The following precautions should be observed to prevent injuries:
   a. Never attempt to insert glass tubing that has a jagged end. All glass tubing should be fire polished before it is inserted into a rubber stopper.
   b. Use water or glycerin as a lubricant on the end of the glass tubing before inserting it into a rubber stopper. Ask your teacher for the proper lubricant. **CAUTION Protect your hands and fingers when inserting glass tubing into a rubber stopper.**
   c. Wear leather gloves or place folded cloth pads between your hands and the glass tubing. Hold the glass tubing as close as possible to the part where it is to enter the rubber stopper. Always point the glass tubing away from the palm of your hand that holds the stopper, as shown in Figure B on the next page. Using a twisting motion, gently push the tubing into the stopper hole.
   d. At the end of the experiment, put on leather gloves or place folded cloth pads between your hands and the glass tubing and remove the rubber stoppers from the tubing to keep them from sticking or "freezing" to the glass. Use a lubricant as directed in step **1b** if the stopper or tubing won't budge.

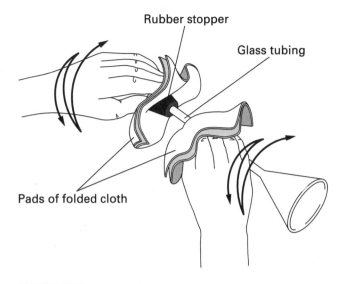

Rubber stopper

Glass tubing

Pads of folded cloth

FIGURE B

2. When inserting glass tubing into a rubber or plastic hose, observe the same precautions discussed in steps **1a–1c.** The glass tubing should be lubricated before insertion into the rubber or plastic hose. The rubber hose should be cut at an angle before the insertion of the glass tubing. The angled cut in the hose allows the rubber to stretch more readily.
CAUTION Protect your hands when inserting or removing glass tubing.
     At the end of an experiment, immediately remove the glass tubing from the hose. When disassembling, follow the precautions that were given in step **1d.**
     Carefully follow these precautions and techniques whenever an experiment requires that you insert glass tubing into either a rubber stopper or a rubber or plastic hose. You will be referred to these safety precautions throughout the lab course.

MATERIALS

PART 3 HANDLING SOLIDS

- glazed paper
- sodium chloride
- spatula
- test tube

PROCEDURE

1. Solids are usually kept in wide-mouthed bottles. A spatula should be used to dip out the solid as shown in Figure C.

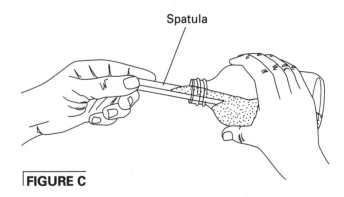

Spatula

FIGURE C

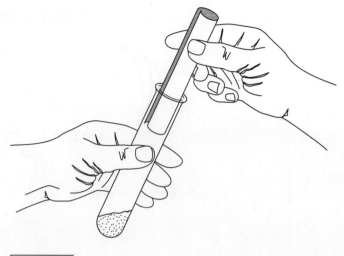

FIGURE D

CAUTION Do not touch chemicals with your hands. Some chemical reagents readily pass through the skin into the bloodstream and can cause serious health problems. Some chemicals are corrosive. Always wear an apron and safety goggles when handling chemicals. Carefully check the label on the reagent bottle or container before removing any of the contents. Never use more of a chemical than directed. You should also know the locations of the emergency lab shower and eyewash station and the procedure for using them in case of an accident.

Using a spatula remove a quantity of sodium chloride from its reagent bottle. In order to transfer the sodium chloride to a test tube, first place it on a piece of glazed paper about 10 cm square. Roll the paper into a cylinder and slide it into a test tube that is lying flat on the table. When you lift the tube to a vertical position and tap the paper gently, the solid will slide down into the test tube, as shown in Figure D.

CAUTION Never try to pour a solid from a bottle into a test tube. As a precaution against contamination, never pour unused chemicals back into their reagent bottles.

2. Dispose of the solid sodium chloride and glazed paper in the waste jars or containers provided by your teacher.

CAUTION Never discard chemicals or broken glassware in the wastepaper basket. This is an important safety precaution against fires, and it prevents personal injuries (such as hand cuts) to anyone who empties the wastepaper basket.

 3. Remember to clean up the lab station and wash your hands at the end of this part of the experiment.

MATERIALS

PART 4 THE BALANCE

- balance, centigram
- glazed paper
- sodium chloride
- spatula
- weighing paper

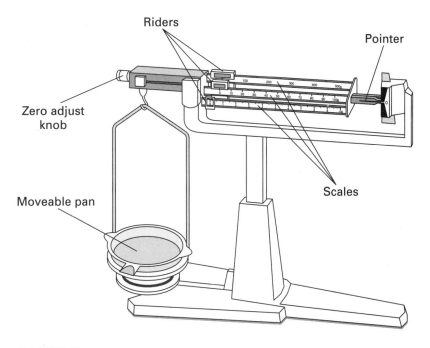

Riders

Pointer

Zero adjust
knob

Scales

Moveable pan

FIGURE E

PROCEDURE

1. When a balance is required for determining mass, you will use a centigram balance like the one shown in Figure E. The centigram balance has a readability of 0.01 g. This means that your mass readings should all be recorded to the nearest 0.01 g.

2. Before using the balance, always check to see if the pointer is resting at zero. If the pointer is not at zero, check the riders on the scales. If all the scale riders are at zero, turn the zero adjust knob until the pointer rests at zero. The zero adjust knob is usually located at the far left end of the balance beam as shown in Figure E. Note: The balance will not adjust to zero if the movable pan has been removed. **Never place chemicals or hot objects directly on the balance pan.** Always use weighing paper or a glass container. Chemicals can permanently damage the surface of the balance pan and affect the accuracy of measurements.

3. In many experiments you will be asked to determine the mass of a specified amount of a chemical solid. Use the following procedure to obtain approximately 13 grams of sodium chloride.
   a. Make sure the pointer on the balance is set at zero. Obtain a piece of weighing paper and place it on the balance pan. Determine the mass of the paper by adjusting the riders on the various scales. Record the mass of the weighing paper to the nearest 0.01 g.

   Mass of paper: Student answer

   b. Add 13 grams to the balance by sliding the rider on the 100 g scale to 10 and the rider on the 10 g scale to 3.
   c. Using a spatula, obtain a quantity of sodium chloride from the reagent bottle and place it on a separate piece of glazed paper.

d. Now slowly pour the sodium chloride from the glazed paper onto the weighing paper on the balance pan, until the pointer once again comes to zero. In most cases, you will only have to be close to the specified mass. Do not waste time trying to obtain exactly 13.00 g. Instead, read the exact mass when the pointer rests close to zero and you have around 13 g of sodium chloride in the balance pan. The mass might be 13.18 g. Record your exact mass of sodium chloride and the weighing paper to the nearest 0.01 g. (Hint: Remember to subtract the mass of the weighing paper to find the mass of sodium chloride.)

Mass of NaCl and paper: Student answer _____

4. Wash your hands thoroughly with soap and water at the end of each lab period.

MATERIALS

PART 5 MEASURING LIQUIDS

- 50 mL beaker
- 250 mL beaker
- 100 mL graduated cylinder
- buret
- buret clamp
- funnel
- pipet
- ring stand
- water

PROCEDURE

1. For approximate measurements of liquids, a graduated cylinder such as the one shown in Figure F is generally used. These cylinders are usually graduated in milliliters (mL). They may also have a second column of graduations reading from top to bottom. Examine your cylinder for these markings. Record the capacity and describe the scale of your cylinder in the space below.

Observation:

_____

_____

_____

2. A pipet or a buret is used for more accurate volume measurements. Pipets are made in many sizes and are used to deliver measured volumes of liquids. A pipet is fitted with a suction bulb as shown in Figure G on the next page. The bulb is used to withdraw air from the pipet while drawing up the liquid to be measured. **CAUTION Always use the suction bulb. NEVER pipet by mouth.**

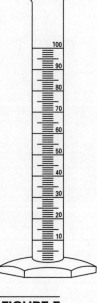

FIGURE F

ChemFile

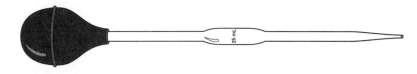

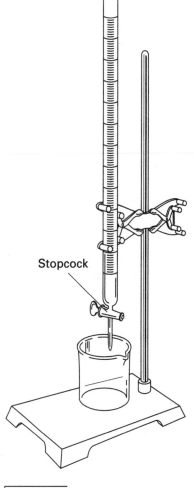

¯FIGURE G

3. Burets are used for delivering any desired quantity of liquid up to the capacity of the buret. Many burets are graduated in tenths of milliliters. When using a buret, follow these steps:

a. Clamp the buret in position on a ring stand as shown in Figure H.

b. Place a 250 mL beaker under the tip of the buret. The beaker serves to catch any liquid that is released.

c. Pour a quantity of liquid you want to measure from the liquid's reagent bottle into a 50 mL beaker. (NOTE: In this first trial you will use water.) Use a glass funnel in the top of the buret to avoid spills when pouring the liquid from the beaker. Carefully check the label of the reagent bottle before removing any liquid. **CAUTION Never pour a liquid directly from its reagent bottle into the buret. You should first pour the liquid into a small, clean, and dry beaker (50 mL) that is easy to handle. Then pour the liquid from the small beaker into the buret. This simple method will prevent unnecessary spillage. Never pour any unused liquid back into the reagent bottle.**

d. Fill the buret with the liquid and then open the stopcock to release enough liquid to fill the tip below the stopcock and bring the level of the liquid within the scale. The height at which the liquid stands is then read accurately. Practice this procedure several times by pouring water into the buret and emptying it through the stopcock.

Stopcock

¯FIGURE H

4. Notice that the surface of a liquid in the buret is slightly curved. It is concave if it wets the glass and convex if it does not wet the glass. Such a curved surface is called a meniscus. If a liquid wets the glass, read the bottom of the meniscus, as shown in Figure I on the next page. This is the line CB. If you read the markings at the top of the meniscus, AD, you will get an incorrect reading. Locate the bottom of the meniscus and read the water level in your buret.

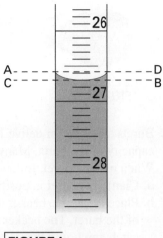

FIGURE I

Buret reading: Student answer _____

5. After you have taken your first buret reading, open the stopcock to release some of the liquid. Then read the buret again. The exact amount released is equal to the difference between your first and final buret reading. Practice measuring liquids by measuring 10 mL of water, using a graduated cylinder, a pipet, and a buret.

6. At the end of this part of the experiment, the equipment you store in the lab locker or drawer should be clean, dry, and arranged in an orderly fashion for the next lab experiment.
CAUTION In many experiments you will have to dispose of a liquid chemical at the end of a lab. Always ask your teacher for the correct method of disposal. In many instances liquid chemicals can be washed down the sink's drain by diluting them with plenty of tap water. Toxic chemicals should be handled only by your teacher. All apparatus should be washed, rinsed, and dried.

7. Remember to wash your hands thoroughly with soap at the end of this part of the experiment.

MATERIALS

PART 6 FILTRATION

- 250 mL beakers, 2
- Bunsen burner with related equipment
- evaporating dish
- filter paper
- fine sand
- funnel
- glass stirring rod

- iron ring
- ring stand
- sodium chloride
- sparker
- wash bottle
- water
- wire gauze, ceramic-centered

PROCEDURE

1. Sometimes liquids contain particles of insoluble solids that are present either as impurities or as precipitates formed by the interaction of the chemicals used in the experiment. If the particles are denser than water, they soon sink to the bottom. Most of the clear, supernatant liquid above the solid may be poured off without disturbing the precipitate. This method of separation is known as decantation.

2. Fine particles or particles that settle slowly are often separated from a liquid by filtration. Support a funnel on a small ring on the ring stand as shown in Figure J. Use a beaker to collect the filtrate. Adjust the funnel so that the stem of the funnel just touches the inside wall of the beaker.

3. Fold a circular piece of filter paper along its diameter, and then fold it again to form a quadrant, as shown in Figure K. Separate the folds of the filter paper, with three thickness on one side and one on the other; then place the filter paper cone in the funnel.

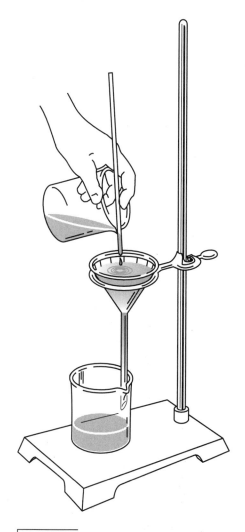

FIGURE J

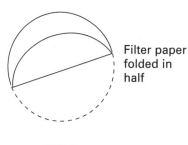

Filter paper folded in half

Filter paper folded in quarters

Filter paper ready for funnel

Filter paper in funnel

FIGURE K

The funnel should be wet before you insert the filter paper. Use your plastic wash bottle to wet the funnel and the filter paper. Press the edges of the filter paper firmly against the sides of the funnel so no air can get between the funnel and the filter paper while the liquid is being filtered. *EXCEPTION: A filter should not be wet with water when the liquid to be filtered does not mix with water. Why?*

Because the liquid does not mix with water, a filter paper wet

with water will not allow the liquid to go through the filter paper.

_____

_____

_____

_____

_____

4. Dissolve 2 or 3 g of salt in a beaker containing about 50 mL of water and stir into the solution an equal volume of fine sand. Filter out the sand by pouring the mixture into the filter, observing the following suggestions:
   a. The filter paper should not extend above the edge of the funnel. Use filter paper that leaves about 1 cm of the funnel exposed.
   b. Do not completely fill the funnel. It must never overflow.
   c. Try to establish a water column in the stem of the funnel to eliminate air bubbles, and then add the liquid quickly enough to keep the mixture level about 1 cm from the top of the filter paper.
   d. When a liquid is poured from a beaker, it may adhere to the glass and run down the outside wall. This may be avoided by holding a stirring rod against the lip of the beaker, as shown in Figure J on the previous page. The liquid will run down the rod and drop off into the funnel without running down the outside of the beaker. The sand is retained on the filter paper. What property of the sand enables it to be separated from the water by filtration?

The sand is insoluble in water.

_____

What does the filtrate contain?

The filtrate contains salt dissolved in water.

_____

5. The salt can be recovered from the filtrate by pouring the filtrate into an evaporating dish and evaporating it over a low flame nearly to dryness. Figure L on the next page shows a correct setup for evaporation.
   CAUTION When using a Bunsen burner, confine loose clothing and long hair. Wear your safety goggles, lab apron, and gloves.

6. Remove the flame as soon as the liquid begins to spatter. Shut off the burner. What property of salt prevents it from being separated from the water by filtration?

Salt is soluble in water.

7. All equipment should be clean, dry, and put away in an orderly fashion for the next lab experiment. Be sure that the valve on the gas jet is completely shut off. Make certain that the filter papers and sand are disposed of in the waste jars or containers and not down the sink. Remember to wash your hands thoroughly with soap at the end of each laboratory period.

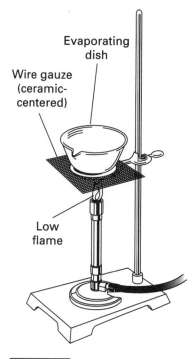

Evaporating dish

Wire gauze (ceramic-centered)

Low flame

FIGURE L

QUESTIONS

Answer the following questions in complete sentences.

1. Organizing Ideas As soon as you enter the lab, what safety equipment should you put on immediately?

Safety goggles and lab apron should be put on.

2. Organizing Ideas Before doing an experiment, what should you read and discuss with your teacher?

All safety precautions in the procedures should be read and

discuss with the teacher.

3. Organizing Ideas Before you light a burner, what safety precautions should always be followed?

Safety goggles and lab apron should be worn. Long hair should

be tied back and loose clothing should be confined. One should

know the locations of fire extinguishers, fire blankets, and sand

buckets and the procedure for using them in case of a fire.

4. **Organizing Ideas** What immediate action should you take when the flame of your burner is burning inside the base of the barrel?

Immediately turn off the gas at the gas valve.

Allow the barrel to cool.

5. **Organizing Ideas** What type of flame is preferred for laboratory work and why?

The nonluminous flame with the light blue inner cone is

preferred because it burns hotter and cleaner than the

yellow luminous flame.

6. **Analyzing Ideas** When inserting glass tubing, why is it important that you wear safety goggles and gloves or cover the tubing and stopper with protective pads of cloth?

These precautions are necessary to prevent personal injuries

such as cuts if the tubing should break. Eyes also need

protection in case the tubing should break and scatter fragments.

7. **Analyzing Ideas** What do you think might be the common cause of fires in lab drawers or lockers?

Putting away hot apparatus before it has had a chance to cool.

8. **Analyzing Ideas** Why are broken glassware, chemicals, matches, and other laboratory debris never discarded in a wastepaper basket?

This is an important safety precaution against fires

and personal injuries, such as cuts from broken glass or chemical

poisoning, to anyone who empties the wastepaper basket.

ChemFile

9. Organizing Ideas List the safety precautions that should be observed when inserting, or removing glass tubing from a rubber stopper or rubber hose.

1. The ends of the glass tubing must be fire polished.

2. A lubricant should be used on the ends of the tubing.

3. Always aim the glass tubing away from the palm of the hand that holds the stopper or hose.

4. Put folded cloth pads between both hands and the glass tubing or wear leather gloves.

5. Use a gentle twisting motion when inserting the tubing.

6. Immediately remove the tubing at the end of an experiment.

7. Follow the same precautions for removal as specified for insertion.

10. Analyzing Ideas Why should you never touch chemicals with your hands?

Some chemicals readily pass through the skin into the bloodstream. Some chemicals are extremely corrosive.

11. Organizing Ideas What precaution can help prevent chemical contamination in reagent bottles?

Never return unused chemicals to reagent bottles.

12. Analyzing Ideas Why are chemicals and hot objects never placed directly on the balance pan?

They can permanently damage the surface of the balance pan and affect the accuracy of mass measurements.

13. Organizing Ideas List three pieces of equipment used in the laboratory for measuring small quantities of liquids. What is the correct procedure for filling a buret with liquid?

Graduated cylinders, pipets, and burets are used to measure liq-

uids. To fill a buret, first pour the liquid into a small beaker that is

easy to handle. Then pour the liquid from the small beaker into

the buret via a glass funnel.

14. Organizing Ideas What is the rule about the size of filter paper to be used with a funnel?

The filter paper should not extend above the edge of the funnel.

Use filter paper that leaves about 1 cm of the funnel exposed

above the edge of the filter cone.

15. Organizing Ideas How can a liquid be transferred from a beaker to a funnel without spattering and without running down the outside wall of the beaker?

A stirring rod may be held against the lip of the beaker. When you

pour, the liquid will cling to the rod, run down it, and drop into

the funnel.

16. Organizing Ideas Describe the condition of all lab equipment at the end of an experiment. What should be checked before you leave the lab?

All equipment that is stored in the lab locker or drawer should be

completely cool, clean, dry, and arranged in an orderly fashion for

the next lab experiment. Check to see that the valve on the gas

jet is completely turned off.

17. Organizing Ideas What is the correct procedure for removing a solid reagent from its container in preparation for use in an experiment?

First, read the label on the reagent bottle before removing any of its contents. Use a clean spatula to remove the approximate quantity needed. Place the reagent on a piece of glazed paper or in a clean, premassed container, as directed by your teacher. Never touch the chemical with your hands and never pour the chemical from its original bottle into another container. Dispose of any unused chemical as directed by your teacher. Never return it to its original container.

18. Organizing Ideas What is the correct procedure for removing a liquid reagent from its container in preparation for use in an experiment?

First, read the label on the reagent bottle before removing any of its contents. When removing the glass stopper from a reagent bottle, never place the stopper on the lab table. Carefully pour the approximate amount of chemical needed into a small, clean, and dry beaker (50 mL) that is easy to handle. Then carefully transfer the liquid into the appropriate devise to measure the volume needed. When transferring to a buret, a funnel may be necessary to avoid spillage. The same cautions for using a solid reagent also apply to liquids.

19. Analyzing Ideas Why is it important to use low flame when evaporating water from a recovered filtrate?

A low flame will heat the evaporating dish and its contents gently. This will prevent the evaporating dish from cracking and will also keep the recovered solid from burning,

GENERAL CONCLUSIONS

Safety Check

Identify the following safety symbols:

a. eye safety

b. heating and fire safety

c. gas precaution

d. clothing protection

e. chemical safety

f. glassware safety

g. hand safety

h. explosion danger

i. radiation precaution

j. caustic substances

Labeling

Practice labeling a chemical container or bottle by filling in the appropriate information missing on the label pictured on the following page. Use 6 M sodium hydroxide (NaOH) as the solution to be labeled. (Hint: 6 M sodium hydroxide is a caustic and corrosive solution and can be considered as potentially hazardous as 6 M HCl.

SAMPLE LABEL

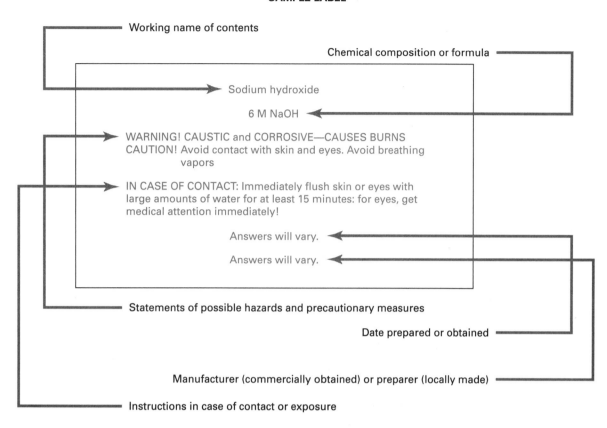

True or False

Read the following statements and indicate whether they are true or false. Place your answer in the space next to the statement.

_____True_____	**1.** Never work alone in the laboratory.
_____True_____	**2.** Never lay the stopper of a reagent bottle on the lab table.
_____False_____	**3.** At the end of an experiment, in order to save the school's money, save all excess chemicals and pour them into their stock bottles.
_____False_____	**4.** The quickest and safest way to heat a material in a test tube is by concentrating the flame on the bottom of the test tube.
_____True_____	**5.** Use care in selecting glassware for high-temperature heating. Glassware should be Pyrex or a similar heat-treated type.
_____True_____	**6.** A mortar and pestle should be used for grinding only one substance at a time.
_____False_____	**7.** Safety goggles protect your eyes from particles and chemical injuries. It is completely safe to wear contact lenses under them while performing experiments.
_____True_____	**8.** Never use the wastepaper basket for disposal of chemicals.
_____False_____	**9.** First aid kits may be used by anyone to give emergency treatment after an accident.
_____True_____	**10.** Eyewash and facewash fountains and safety showers should be checked daily for proper operation.

Chemical Apparatus

Identify each piece of apparatus. Place your answers in the spaces provided.

a.  b.  c.   d.

e.  f.   g.   h.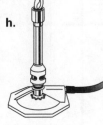

i. 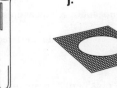 j.   k.   l.

a. beaker

b. graduated cylinder

c. spatula

d. pipestem triangle

e. funnel

f. evaporating dish

g. Erlenmeyer flask

h. Bunsen burner

i. test tube

j. ceramic-centered wire gauze

k. crucible and cover

l. mortar and pestle

Name _____

Date _____ Class _____

EXPERIMENT

Accuracy and Precision in Measurement

OBJECTIVES

Recommended time:
60 min

Required Precautions

• Read all safety cautions and discuss them with your students.

• **Use** experimental measurements in calculations.

• **Organize** data by compiling it in tables.

• **Compute** an average value from class data and use the value to calculate absolute deviation and average deviation.

• **Recognize** the importance of accuracy and precision in scientific measurements.

• **Relate** the reliability of experimental data to absolute deviation, average deviation, uncertainty, and percent error.

INTRODUCTION

• Safety goggles and a lab apron must be worn at all times.

• Tie back loose hair and long clothing when working in the lab.

Procedural Tips

• Students who have done Experiment 1-1 will already know how to read volume gradations on a graduated cylinder and use the balance. They may only need help in reading more precisely.

In this experiment, you will determine the volume of a liquid in two different ways and compare the results. You will also calculate the density of a metal using your measurements of its mass and volume.

$$D = \frac{m}{V}$$

You will compare your result with the accepted value found in a handbook. The error and percent error in each part of the experiment will be calculated.

The *experimental error* is calculated by subtracting the accepted value from the observed or experimental value. The *percent error* is calculated according to the following equation.

$$\text{Percent error} = \frac{Observed\ value\ -\ Accepted\ value}{Accepted\ value} \times 100$$

The sign of the experimental error and the percent error may be either positive (the experimental result is too high) or negative (the experimental result is too low).

You will average the values for the density of a metal obtained by the entire class to determine the average value. Using this value you will calculate the *average deviation* of these data. The average deviation will be expressed as the *uncertainty* of the measurements.

SAFETY

Always wear safety goggles and a lab apron to protect your eyes and clothing. If you get a chemical in your eyes, immediately flush the chemical out at the eyewash station while calling to your teacher. Know the location of the emergency lab shower and eyewash station and the procedure for using them.

Do not touch any chemicals. If you get a chemical on your skin or clothing, wash the chemical off at the sink while calling to your teacher. Make sure you carefully read the labels and follow the precautions on all containers of chemicals that you use. If there are no precautions stated on the label, ask your teacher what precautions to follow. Never return leftover chemicals to their original containers; take only small amounts to avoid wasting supplies.

 Never put broken glass into a regular waste container. Broken glass should be disposed of separately according to your teacher's instructions.

MATERIALS

- 15 cm plastic ruler
- 25 mL graduated cylinder
- 100 mL beaker
- 100 mL graduated cylinder
- balance
- metal shot (aluminum, copper, lead)
- thermometer, nonmercury, 0–100°C

PROCEDURE

- Students may have difficulty understanding the difference between accuracy and precision. You might use the analogy of the dart board. If darts puncture the board over a large circular area surrounding the bull's eye, the shots may be considered to be accurate because an average of the positions of all the shots would come close to the position of the bull's eye. However, the shots would not be precise because they are not close together. On the other hand, a distribution of shots such that they are clustered in a small area below the bull's eye might be considered precise because they are close together, but not accurate because their positions do not average to that of the bull's eye.

RECORDING YOUR OBSERVATIONS

After completing each part of the experiment, record your observations in the appropriate data table. Recording data anywhere else increases the probability of recording an inaccurate value.

Part 1

1. Examine the centimeter scale of the plastic ruler. What are the smallest divisions?

The smallest divisions on the ruler are millimeters.

To what fraction of a centimeter can you make measurements with such a ruler?

You should be able to measure to ± 0.05 cm.

2. Using the ruler, measure the inside diameter of the 100 mL graduated cylinder. Similarly, measure the inside height of the cylinder to the 50 mL mark. Record these measurements in Data Table 1.

Data Table 1	
Inside diameter of graduated cylinder	2.50 cm
Inside height of graduated cylinder	9.31 cm

Part 2

3. Examine the gram scale of the balance. What are the smallest divisions?

The smallest divisions are centigrams.

To what fraction of a gram can you make measurements with a centigram balance?

You should be able to measure to $\pm$ 0.01 g.

4. Examine the graduations on a 25 mL graduated cylinder and determine the smallest fraction of a milliliter to which you could make a measurement. Does this match the uncertainty of a measurement made with a 100 mL graduated cylinder?

Most 25 mL cylinders are subdivided by 0.5-mL

gradations, whereas 100 mL cylinders are subdivided

by 1-mL gradations. The uncertainties of measurements

using the two cylinders do not match ($\pm$ 0.25 mL for the

25 mL cylinder and $\pm$ 0.5 mL for the 100 mL cylinder).

5. Using the balance, determine the mass of the dry 25 mL cylinder. Record the mass in Data Table 2.

6. Fill the beaker half full of water and determine its temperature to the nearest degree. Look up the density of water for this temperature, and record in the data table both the temperature and water density.

7. Fill your graduated cylinder with water to a level between 10 and 25 mL; accurately read and record the volume. Determine the mass of the water plus the cylinder. Then, record this value in Data Table 2. Save the water in the graduated cylinder for use in Part 3.

Data Table 2		
Mass of empty graduated cylinder	42.39	g
Temperature of water	25.0	°C
Density of water	0.997	g/cm³
Volume of water	13.80	mL
Mass of graduated cylinder + water	55.98	g

Part 3

8. Add a sufficient quantity of the assigned metal shot (aluminum, copper, or lead) to the cylinder containing the water (saved from Part 2) to increase the volume by at least 5 mL. Determine the volume and then the mass of the shot, water, and cylinder. Record your measurements in Data Table 3.

Data Table 3		
Volume of water (from Part 2)	13.80	mL
Mass of water + graduated cylinder (from Step 2)	55.98	g
Volume of metal and water	20.25	mL
Mass of metal + water + graduated cylinder	73.98	g

Cleanup and Disposal

9. Clean up all apparatus and your lab station. Return equipment to its proper place. Dispose of chemicals and solutions in the containers designated by your teacher. Do not pour any chemicals down the drain or in the trash unless your teacher directs you to do so. Wash your hands thoroughly with soap before you leave the lab and after all work is finished.

CALCULATIONS

Show all your calculations. Place your answers in the appropriate calculations table.

Part 1

1. Organizing Data Calculate the volume of the cylinder to the 50.0-mL graduation ($V = 3.14 \times r^2 \times h$).

$$V = 3.14\, r^2 h = 3.14 \times \left(\frac{2.50\text{ cm}}{2}\right)^2 \times 9.31\text{ cm} = 45.7\text{ cm}^3$$

2. Inferring Conclusions Assume the accepted value is 50.0 cm³. Calculate the error and percent error.

Error: 45.7 cm³ − 50.0 cm³ = 4.3 cm³

Percent error: $\dfrac{4.3\text{ cm}^3}{50.0\text{ cm}^3} \times 100 = 8.6\%$

Part 2

3. Organizing Data Calculate the mass of water as measured by the balance.

(Mass of cylinder + water) − mass of cylinder = mass of water

55.98 g − 42.39 g = 13.59 g of water

4. **Organizing Data** Calculate the mass of the water from its measured volume and its density ($m = D \times V$).

$m = D \times V = 0.997 \text{ g/cm}^3 \times 13.80 \text{ cm}^3 = 13.8 \text{ g}$

5. **Inferring Conclusions** Using the mass of water determined by the use of the balance as the *accepted value,* calculate the error and percent error in the mass determined using the volume and density.

Error: 13.80 g − 13.59 g = 0.21 g

Percent error: $\dfrac{0.21 \text{ g}}{13.59 \text{ g}} \times 100 = 1.5\%$

Part 3

6. **Organizing Data** Determine the volume of the metal using your measurement of the volume of water displaced by the metal.

Volume of metal = 20.25 mL − 13.80 mL = 6.45 mL = 6.45 cm³

7. **Organizing Data** Using your measurements in Data Table 3, determine the mass of the metal.

Mass of metal = 73.98 g − 55.98 g = 18.00 g

8. **Organizing Data** Calculate the density of the metal.

Density of metal = $\dfrac{18.00 \text{ g}}{6.45 \text{ cm}^3} = 2.79 \text{ g/cm}^3$

9. **Inferring Conclusions** In a handbook or the appendix of your textbook, look up the specific gravity of the metal you used. (In the metric system, the density of liquids and solids is equal to the specific gravity.) Calculate the error and percent error for the density of the metal shot you determined in item **8.**

Experimental value − accepted value = error

Error: 2.79 g/cm³ − 2.7 g/cm³ = 0.1 g/cm³

Percent error: $\dfrac{0.1 \text{ g/cm}^3}{2.7 \text{ g/cm}^3} \times 100 = 4\%$

Part 4

10. Organizing Conclusions   Record in the table below five values obtained by you and your classmates for the density of the same metal.

Group Number	Density (g/cm³)
1	2.79
2	3.00
3	2.95
4	2.79
5	2.81

11. Evaluating Conclusions   Calculate the average density (M) of the five results.

$$M = \frac{2.79 + 3.00 + 2.95 + 2.79 + 2.81}{5} = 2.87 \text{ g/cm}^3$$

QUESTIONS

1. Evaluating Methods   What value of a measurement must be known if the accuracy of an experimental measurement is to be determined?

The accepted value of a measurement must be known.

_____

_____

_____

2. Evaluating Methods   What are the possible sources of experimental errors in this experiment?

Sources of experimental errors include inaccurate

observations and imperfections in apparatus design.

_____

_____

GENERAL CONCLUSIONS

1. Evaluating Conclusions   Sarah and Jamal determined the density of a liquid three times. The values they obtained were 2.84 g/cm³, 2.85 g/cm³, and 2.80 g/cm³. The accepted value is known to be 2.40 g/cm³.
a. Are the values that Sarah and Jamal determined precise? Explain.

Yes, the measurements are in close agreement

_____

b. Are the values accurate? Explain.

No, they are not close enough to the real value.

c. Calculate the percent error and the uncertainty for each measurement.

Uncertainty: Average Density 2.83 g/cm³

Experiment	Experimental density (g/cm³)	Average density (g/cm³)	Absolute deviation g/cm³
1	2.84	2.83	0.1
2	2.85	2.83	0.2
3	2.80	2.83	0.3

Percent error

1. $\dfrac{2.84 \text{ g/cm}^3 - 2.40 \text{ g/cm}^3}{2.40 \text{ g/cm}^3} \times 100 = 18.3\%$

2. $\dfrac{2.85 \text{ g/cm}^3 - 2.40 \text{ g/cm}^3}{2.40 \text{ g/cm}^3} \times 100 = 18.8\%$

3. $\dfrac{2.80 \text{ g/cm}^3 - 2.40 \text{ g/cm}^3}{2.40 \text{ g/cm}^3} \times 100 = 16.7\%$

EXPERIMENT **A3**

Reactivity of Halide Ions

OBJECTIVES

Recommended time:
60 min

- **Observe** the reactions of halide ions with different reagents.
- **Analyze** the data to determine characteristic reactions of each halide ion.
- **Infer** the identity of unknown solutions based on your observations of known reactions.

INTRODUCTION

Solution/Material Preparation

1. To prepare 500 mL of 0.1 M $AgNO_3$ solution, dissolve 8.5 g of $AgNO_3$ in enough distilled water to make 500 mL of solution.

2. To prepare 1 L of 0.1 M NaCl solution, dissolve 6 g of NaCl in enough distilled water to make 1 L of solution.

3. To prepare 1 L of 0.1 M NaF solution, dissolve 4 g of NaF in enough distilled water to make 1 L of solution.

The four halide salts used in this experiment are found in your body. Although sodium fluoride is poisonous, trace amounts seem to help strengthen tooth enamel and prevent tooth decay in humans. Sodium chloride is added to most of our food to enhance flavor while masking sourness and bitterness. Sodium chloride is essential for life processes, but excessive intake of this salt appears to be linked to high blood pressure in certain segments of the population. Sodium bromide is distributed throughout body tissues, and in the past it was used as a sedative. Sodium iodide is necessary for the proper function of the thyroid gland, which controls cell growth. The concentration of sodium iodide is almost 20 times greater in the thyroid than in blood. To help meet the body's need for this halide salt, about 10 ppm of NaI is added to packages of table salt that are labeled "iodized."

The principal oxidation state of the halogens is -1. However, all halogens except fluorine may exist in other oxidation states. The specific tests you will develop in this experiment involve the production of recognizable precipitates and complex ions. You will use your observations to determine the halide ion present in an unknown solution.

SAFETY

4. To prepare 1 L of 0.2 M KBr solution, dissolve 24 g of KBr in enough distilled water to make 1 L of solution.

5. To prepare 1 L of 0.2 M KI, dissolve 33 g of KI in enough distilled water to make 1 L of solution.

6. To prepare 1 L of 0.2 M $Na_2S_2O_3$ solution, dissolve 50 g of $Na_2S_2O_3 \cdot 5H_2O$ in enough distilled water to make 1 L of solution.

Always wear safety goggles, disposable gloves, and a lab apron to protect your eyes, hands, and clothing. If you get a chemical in your eyes, immediately flush the chemical out at the eyewash station while calling to your teacher. Know the location of the emergency lab shower and eyewash station and the procedure for using them.

Do not touch any chemicals. If you get a chemical on your skin or clothing, wash the chemical off at the sink while calling to your teacher. Make sure you carefully read the labels and follow the precautions on all containers of chemicals that you use. If there are no precautions stated on the label, ask your teacher what precautions you should follow. Do not taste any chemicals or items used in the laboratory. Never return leftover chemicals to their original containers; take only small amounts to avoid wasting supplies.

7. To prepare 500 mL of 0.5 M Ca(NO₃)₂ solution, dissolve 41 g of Ca(NO₃)₂ in enough distilled water to make 500 mL of solution.

 Call your teacher in the event of a spill. Spills should be cleaned up promptly, according to your teacher's directions.

 Never put broken glass into a regular waste container. Broken glass should be disposed of properly in the broken glass waste container.

MATERIALS

8. To prepare 3% starch solution, add 30 g of soluble starch to approximately 800 mL of water. Boil to dissolve the starch, and then add enough distilled water to make 1 L of solution.

- 0.1 M AgNO₃
- 0.1 M NaCl
- 0.1 M NaF
- 0.2 M KBr
- 0.2 M KI
- 0.2 M Na₂S₂O₃
- 0.5 M Ca(NO₃)₂
- 3% starch solution
- 5% NaOCl (commercial bleach)
- 10 mL graduated cylinder
- test tubes, 18 mm × 150 mm, 12

PROCEDURE

9. If a mixture of two halide ions is used for an unknown, double the concentration of each original solution.

Required Precautions

- Read all safety precautions, and discuss them with your students.
- Safety goggles, disposable gloves, and a lab apron must be worn at all times.
- In case of a spill, use a dampened cloth or paper towels to mop up the spill. Then rinse the towel in running water at the sink, wring it out until it is only damp and put it in the trash.

1. Place 5 mL of NaF, NaCl, KBr, and KI solutions into separate test tubes. Label the test tubes to identify their contents. Add 1 mL of 0.5 M Ca(NO₃)₂ to each test tube. Record the color of any precipitates that form in the Data Table on the next page. Indicate which, if any, did not form a precipitate.

2. Place 5 mL each of NaF, NaCl, KBr, and KI solutions into another set of clean, labeled test tubes. Add 1 mL of 0.1 M AgNO₃ to each of the test tubes. Record your observations in your data table.

3. Add 5 mL of 0.2 M Na₂S₂O₃ solution to each test tube in the set of test tubes containing precipitates from Step **2**. Record your observations.

4. Place 2 mL each of NaF, NaCl, KBr, and KI solutions into separate, clean, labeled test tubes. Add 5 mL of 3% starch solution to each test tube. Add a drop or two of 5% NaOCl (commercial bleach) to each test tube. Record your observations.

5. Save the results of the tests on the known samples for comparison, and obtain an unknown solution containing one halide ion. React the unknown with each of the reagents as you did in Steps **1–4**. Record your observations in your data table.

6. Optional: Obtain an unknown solution containing a mixture of two halide ions. Repeat Steps **1–4**. Record your observations.

Cleanup and Disposal

 7. Clean all apparatus and your lab station. Return equipment to its proper place. Dispose of chemicals and solutions in the containers designated by your teacher. Do not pour any chemicals down the drain or in the trash unless your teacher directs you to do so. Wash your hands thoroughly before you leave the lab and after all work is finished.

Halide salts	Ca(NO₃)₂	AgNO₃	AgNO₃ + Na₂S₂O₃	NaOCl + starch
NaF	white ppt	clear	no change	no change
NaCl	clear	white ppt	dissolves completely	no change
KBr	clear	pale-yellow ppt	dissolves completely	no change
KI	clear	yellow-tan ppt	does not dissolve	turns blue
Single unknown	student observations	student observations	student observations	student observations
Double unknown	student observations	student observations	student observations	student observations

Data Table—Reagents

QUESTIONS

Disposal

Combine all solutions. Add sufficient 0.2 M KI solution to precipitate all silver as AgI. Then dilute 100-fold and pour down the drain.

1. **Analyzing Data** Which step(s) confirm(s) the presence of (a) F⁻ ions, (b) Cl⁻ ions, (c) Br⁻ ions, (d) I⁻ ions?

   a. Step 1 _____

   b. Step 2 _____

   c. Steps 2 and 3 _____

   d. Step 4 _____

2. **Inferring Conclusions** What generalizations can be made about silver halides?

   They are all insoluble except for NaF. _____

   _____

GENERAL CONCLUSIONS

1. **Inferring Conclusions** Identify your unknown(s), and use your experimental evidence to support your identifications.

   Student answers will vary. _____

   _____

   _____

Name _____

Date _____ Class _____

HOLT
ChemFile
LAB PROGRAM

Tests for Iron(II) and Iron(III)

OBJECTIVES

Recommended time:
60 min

- **Observe** tests of known solutions containing iron(II) or iron(III) ions.
- **Compare** results for the two ions, and infer conclusions.
- **Design** a procedure for identifying the two ions in one solution.

INTRODUCTION

Solution/Material Preparation

1. Wear safety goggles, a face shield, impermeable gloves, and a lab apron when you prepare the iron(III) chloride solution containing HCl. Work in a hood known to be in operating condition and have another person stand by to call for help in case of an emergency. Be sure you are within a 30 s walk from a safety shower and eyewash station known to be in good operating condition.

2. In preparing the solutions, care should be taken that the solid potassium ferrocyanide and potassium ferricyanide are not heated and do not come into contact with concentrated acids.

In this experiment, the complex hexacyanoferrate(II) ion (*ferro*cyanide), $Fe(CN)_6^{4-}$, and the hexacyanoferrate(III) ion (*ferri*cyanide), $Fe(CN)_6^{3-}$, will be used in identification tests for Fe^{2+} and Fe^{3+} ions. The charges on the two complex ions clearly indicate the difference in the oxidation state of the iron present in each. The (CN) group in each complex ion has a charge of -1. Thus, iron(II) is present in the *ferro*cyanide ion, $[Fe^{2+}(CN^-)_6]^{4-}$. Iron(III) is present in the *ferri*cyanide ion group, $[Fe^{3+}(CN^-)_6]^{3-}$. A deep-blue precipitate results when either complex ion combines with iron in a different oxidation state from that present in the complex. The deep-blue color of the precipitate is caused by the presence of iron in both oxidation states. The color provides a means of identifying either iron ion. If the deep-blue precipitate is formed on addition of the $[Fe^{2+}(CN^-)_6]^{4-}$ complex, the iron ion responsible must be the iron(III) ion. Similarly, a deep-blue precipitate formed with the $[Fe^{3+}(CN^-)_6]^{3-}$ complex indicates the presence of the iron(II) ion.

Both of the deep-blue precipitates are known to have the same composition. The potassium salt of the complex ion has the formula $KFeFe(CN)_6 \cdot H_2O$.

The thiocyanate ion, SCN^-, provides a test for confirming the presence of Fe^{3+} ion. The soluble $FeSCN^{2+}$ complex is formed, imparting a blood-red color to the solution.

The deep-blue precipitate produced in this experiment is found in a number of products. This insoluble $KFeFe(CN)_6 \cdot H_2O$ complex is the "blue" of blueprint paper. It is the pigment in Prussian blue oil paint used by artists for over 300 years. It is used in inks, in carbon paper, and as the main ingredient in certain brands of laundry bluing used to counteract the yellowing of white clothes.

SAFETY

Always wear safety goggles, disposable gloves, and a lab apron to protect your eyes, hands, and clothing. If you get a chemical in your eyes, immediately flush the chemical out at the eyewash station while calling to your teacher. Know the location of the emergency lab shower and eyewash station and the procedure for using them.

3. To prepare 1 L of 0.1 M iron(III) chloride, observe the required safety precautions. Add 20 mL of concentrated HCl, with stirring, to approximately 800 mL of distilled water. Dissolve 27 g of $FeCl_3 \cdot 6H_2O$ in the acidified water and then dilute to 1 L with distilled water.

4. To prepare 500 mL of 0.1 M potassium hexacyanoferrate(III), dissolve 17 g of $K_3Fe(CN)_6$ in enough distilled water to make 500 mL of solution.

Do not touch any chemicals. If you get a chemical on your skin or clothing, wash the chemical off at the sink while calling to your teacher. Make sure you carefully read the labels and follow the precautions on all containers of chemicals that you use. If there are no precautions stated on the label, ask your teacher what precautions you should follow. Do not taste any chemicals or items used in the laboratory. Never return leftover chemicals to their original containers; take only small amounts to avoid wasting supplies.

Call your teacher in the event of a spill. Spills should be cleaned up promptly, according to your teacher's directions.

Never put broken glass in a regular waste container. Broken glass should be disposed of properly in the broken glass waste container.

MATERIALS

5. To prepare 500 mL of 0.1 M potassium hexacyanoferrate(II), dissolve 21 g of $K_4Fe(CN)_6 \cdot 3H_2O$ in enough distilled water to make 500 mL of solution.

- 0.1 M $FeCl_3$
- 0.1 M $K_3Fe(CN)_6$
- 0.1 M $K_4Fe(CN)_6$
- 0.2 M $Fe(NH_4)_2(SO_4)_2$
- 0.2 M KSCN
- 10 mL graduated cylinder
- 150 mL beaker
- glass stirring rod
- test tubes, 18 mm × 150 mm, 6

PROCEDURE

6. The iron(II) ammonium sulfate solution must be freshly prepared because it will oxidize. To prepare 100 mL of 0.2 M iron(II) ammonium sulfate, dissolve 7.8 g of $Fe(NH_4)_2(SO_4)_2 \cdot 6H_2O$ in enough distilled water to make 100 mL of solution.

7. To prepare 100 mL of 0.2 M potassium thiocyanate, dissolve 1.9 g KSCN in enough distilled water to make 100 mL of solution.

Required Precautions

- Safety goggles, disposable gloves, and a lab apron must be worn at all times.

1. In a 150 mL beaker, dissolve 3.7 g of iron(II) ammonium sulfate, $Fe(NH_4)_2(SO_4)_2$, in approximately 100 mL of distilled water.

2. Transfer a 5-mL portion of the iron(II) ammonium sulfate solution to a test tube, and add 3 or 4 drops of 0.2 M potassium thiocyanate solution, KSCN. Record your observations in your data table.

3. Transfer another 5-mL portion of the iron(II) ammonium sulfate solution to a test tube, and add 1 mL of 0.1 M $K_3Fe(CN)_6$ solution ($[Fe^{3+}(CN^-)_6]^{3-}$ ion). Record your observations.

4. To a 5-mL portion of iron(III) chloride, $FeCl_3$, add 1 mL of 0.1 M $K_3Fe(CN)_6$ solution. Record your observations in your data table.

5. To a second 5-mL portion of iron(III) chloride solution, add 1 mL of 0.1 M $K_4Fe(CN)_6$ ($[Fe^{2+}(CN^-)_6]^{4-}$ ion). Record your observations in your data table.

6. To a third 5-mL portion of iron(III) chloride solution, add 2 or 3 drops of the KSCN solution. Record your observations.

Data Table

	Hexacyanoferrate (II) ion	Hexacyanoferrate (III) ion	Thiocyanate ion
Iron ion	$[Fe^{2+}(CN^-)_6]^{4-}$	$[Fe^{3+}(CN^-)_6]^{3-}$	SCN^-
Fe^{2+}		dark-blue ppt	no visible result
Fe^{3+}	dark-blue ppt	brown solution	blood-red solution
Observations in step 2	No color or precipitate forms.		
Observations in step 3	A deep-blue precipitate forms.		
Observations in step 4	No precipitate forms, but the solution turns dark brown.		
Observations in step 5	A dark-blue precipitate forms.		
Observations in step 6	The solution turns blood red because of the presence of the $FeSCN^{2+}$ ion.		

• In case of an acid or base spill, first dilute the substance with water. Then, mop up the spill with wet cloths or a wet cloth mop while wearing disposable plastic gloves. Designate separate cloths or mops for acid and base spills.

Cleanup and Disposal

7. Clean all apparatus and your lab station. Return equipment to its proper place. Dispose of chemicals and solutions in the containers designated by your teacher. Do not pour any chemicals down the drain or in the trash unless your teacher directs you to do so. Wash your hands thoroughly before you leave the lab and after all work is finished.

QUESTIONS

1. **Organizing Ideas** Explain specifically how you would make a conclusive test for an iron(III) salt.

   Add $K_4Fe(CN)_6$ to the solution. The formation of a deep-blue

   precipitate identifies the iron(III) ion. Confirm by adding KSCN

   to a second test solution. The formation of a blood-red solution

   confirms the presence of an iron(III) salt.

2. **Organizing Ideas** Which test for iron(II) ions is conclusive?

   The formation of a dark-blue precipitate on addition of $K_3Fe(CN)_6$

   to a solution containing iron(II) ions.

3. **Relating Ideas** When iron(II) ammonium sulfate is mixed with the $[Fe^{2+}(CN^-)_6]^{4-}$ ion, the precipitate is initially white but turns blue upon exposure to air. What is happening to the iron(II) ion when the precipitate turns blue?

Iron(II) is being oxidized in the presence of air to iron(III).

GENERAL CONCLUSIONS

Disposal

Combine all solutions and precipitates. Filter. To the filtrate, add 0.1 M potassium hexa-cyanoferrate(III) solution until no more precipitate forms. Filter. To the filtrate, add 0.1 M potassium hexacynofer-rate(II) solution until no more precipitate forms. Filter. Adjust the pH of the filtrate to be within 5 to 9 by adding 1.0 M NaOH solution, if necessary. Then pour down the drain. Combine all precipitates and put in the trash after drying.

1. **Designing Experiments** Suppose you have a solution containing both an iron(II) salt and an iron(III) salt. How would you proceed to identify both Fe^{2+} and Fe^{3+} in this solution?

$K_3Fe(CN)_6$ could be added to one sample of the solution to

detect the iron(II) ion, and $K_4Fe(CN)_6$ could be added to

another sample of the solution to identify the iron(III) ion.

2. **Relating Ideas** Blueprint paper can be made by soaking paper in a brown solution of $[Fe^{3+}(CN^-)_6]^{3-}$ and iron(III) ammonium citrate. Wherever the paper is exposed to bright light, the paper turns blue. Explain why this happens.

The bright light changes the iron(III) ion to iron(II), which then

precipitates with the ferricyanide as Prussian blue. This deep-blue

compound is insoluble in water.

Name _____

Date _____ Class _____

Evidence for Chemical Change

OBJECTIVES

Recommended time:
60 min

- **Observe** evidence that a chemical change has taken place.
- **Infer** from observations that a new substance has been formed.
- **Identify** and record observations that show heat is involved in chemical change.
- **Observe** the color and solubility of some substances.
- **Describe** reactions by writing word equations.

INTRODUCTION

One way of knowing that a chemical change has occurred is to observe that the properties of the products are different from those of the reactants. The new product can then become a reactant in another chemical reaction. In this experiment you will observe a sequence of changes that occur when a solution that begins as copper(II) nitrate is treated with a series of different reactants. All of the reactions will take place in the same test tube. At each step you will look for evidence that a new substance is formed as a result of a chemical change. You will also observe heat changes and relate them to chemical reactions.

SAFETY

Solution/Material Preparation

1. Wear safety goggles, a face shield, impermeable gloves, and a lab apron when you prepare the 1.0 M HCl and the 1.0 M NaOH. When making the HCl solution, work in a hood known to be in operating condition and have another person stand by to call for help in case of an emergency. Be sure you are within a 30 s walk from a safety shower and eyewash station known to be in good operating condition.

  Always wear safety goggles and a lab apron to protect your eyes and clothing. If you get a chemical in your eyes, immediately flush the chemical out at the eyewash station while calling to your teacher. Know the location of the emergency lab shower and the eyewash station and the procedure for using them.

 Do not touch any chemicals. If you get a chemical on your skin or clothing, wash the chemical off at the sink while calling to your teacher. Make sure you carefully read the labels and follow the precautions on all containers of chemicals that you use. If there are no precautions stated on the label, ask your teacher what precautions you should follow. Do not taste any chemicals or items used in the laboratory. Never return leftover chemicals to their original containers; take only small amounts to avoid wasting supplies.

 Call your teacher in the event of a spill. Spills should be cleaned up promptly, according to your teacher's directions.

2. To prepare 1 L of 1.0 M copper(II) nitrate, dissolve 296 g of $Cu(NO_3)_2 \cdot 6H_2O$ or 242 g of $Cu(NO_3)_2 \cdot 3H_2O$ in enough distilled water to make 1.0 L of solution.

 When using a Bunsen burner, confine long hair and loose clothing. If your clothing catches on fire, WALK to the emergency lab shower, and use it to put out the fire. Do not heat glassware that is broken, chipped, or cracked. Use tongs or a hot mitt to handle heated glassware and other equipment because heated glassware does not look hot.

 Never put broken glass into a regular waste container. Broken glass should be disposed of properly.

MATERIALS

- 1.0 M copper(II) nitrate
- 1.0 M HCl
- 1.0 M NaOH
- 100 mL beaker
- aluminum wire, 12 cm
- Bunsen burner and related equipment
- glass stirring rod

- iron ring
- lab marker
- ring stand
- ruler
- test tube, 13 mm × 100 mm
- test-tube rack
- wire gauze, ceramic-centered

PROCEDURE

3. To prepare 1 L of 0.1 M sodium hydroxide, observe the required safety precautions. Slowly and with stirring, dissolve 40 g of NaOH in enough distilled water to make 1.0 L of solution.

Required Precautions

- Read all safety precautions, and discuss them with your students.

- Safety goggles and a lab apron must be worn at all times.

- In case of a spill, use a dampened cloth or paper towels to mop up the spill. Then, rinse the towel in running water at the sink, wring it out until it is only damp, and put it in the trash. In the event of an acid or base spill, dilute with water first, and then proceed as described.

Procedural Tips

- Thin-stemmed pipets or dropper bottles may be used to dispense the solutions.

1. Place 50 mL of water into the 100 mL beaker and heat it until boiling. This will be the water bath you will use in Step **5**.

2. While the water bath is heating, use the lab marker and ruler to make three marks on the test tube that are 1 cm apart. Make the marks starting at the bottom of the test tube and moving toward the top.

3. Add 1.0 M copper(II) nitrate to the first mark on the test tube, as shown in Figure A.

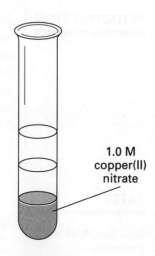

1.0 M copper(II) nitrate

FIGURE A

4. Add 1.0 M sodium hydroxide up to the second mark on the test tube, as shown in Figure B. **CAUTION Sodium hydroxide is corrosive. Be certain to wear safety goggles and a lab apron. Avoid contact with skin and eyes. If any of this solution should spill on you, immediately flush the area with water and then notify your teacher.**
Mix the solutions with the stirring rod. Rinse the stirring rod thoroughly before setting it down on the lab table. Touch the bottom of the outside of the test tube to see if heat has been released. The copper-containing product in the test tube is copper(II) hydroxide. The other product is sodium nitrate. Record the changes that occur in the test tube in the space provided on the next page.

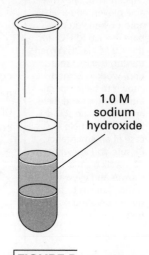

1.0 M sodium hydroxide

FIGURE B

● Emphasize the importance of making the marks on the test tube carefully to insure accurate measurements. You may wish to set up several hot water baths in the room to eliminate student use of the burner.

● Encourage students to make careful observations using all their senses (except taste). Remind them to look for changes that indicate that a new substance has formed, and to record their observations immediately.

Disposal

Combine all solutions and precipitates containing copper. Add 1.0 M NaOH slowly and with stirring until all the copper has precipitated as the hydroxide. Filter. After drying, put the precipitate into the trash. Slowly and with stirring, add all the waste HCl solution to the filtrate. Then adjust the pH of this mixture to be between 5 and 9 by adding 1.0 M acid or base, as necessary. Then pour down the drain.

Observations:

A blue precipitate is formed. Heat is evolved.

5. Put the test tube into the water bath you prepared in Step **1.** Heat it until no more changes occur. The products of this reaction are copper(II) oxide and water. Record the changes that occur in the test tube.

Observations:

The precipitate turns black.

6. Remove the test tube from the hot-water bath. Turn off the burner. Cool the test tube and its contents for 2 min in room-temperature water. Add 1.0 M hydrochloric acid to the third mark, as shown in Figure C.
CAUTION Hydrochloric acid is corrosive. Be certain to wear safety goggles and a lab apron. Avoid contact with skin and eyes. Avoid breathing vapors. If any of this solution should spill on you, immediately flush the area with water and then notify your teacher.
Mix with the stirring rod. Rinse the stirring rod.
The new products are copper(II) chloride and water. Record the changes that occur in the test tube.

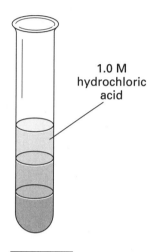

1.0 M hydrochloric acid

FIGURE C

Observations:

The precipitate dissolves and the solution turns green.

7. Place a 12-cm piece of aluminum wire in the test tube. Leave it until no reaction is observed. Touch the bottom of the test tube to check for temperature change. Two reactions take place. Copper(II) chloride and aluminum produce copper and aluminum chloride. The aluminum also reacts with the hydrochloric acid to form hydrogen and aluminum chloride. Record the changes that occur in the test tube.

Observations:

The green color disappears and a dark reddish solid forms

on the wire. Heat is evolved. Bubbles are produced.

8. Remove the wire from the test tube. Compare the copper formed to a sample of copper wire. Record your observations.

Observations:

The color is similar but the shape and form are not.

Cleanup and Disposal

9. Clean all apparatus and your lab station. Return equipment to its proper place. Dispose of chemicals and solutions in the containers designated by your teacher. Do not pour any chemicals down the drain or in the trash unless your teacher directs you to do so. Wash your hands thoroughly before you leave the lab and after all work is finished.

QUESTIONS

1. Organizing Ideas What are some causes of chemical changes?

combining chemicals; adding heat

2. Organizing Ideas In what two ways is heat involved in chemical change? Cite some specific instances from this experiment.

(1) heat initiates chemical change—changing copper(II) hy-

droxide to copper(II) oxide; (2) heat is evolved during chemical

change—copper(II) nitrate + sodium hydroxide and aluminum +

copper(II) chloride

3. Analyzing Information Identify all of the substances that are used or produced in this experiment. Distinguish between elements and compounds.

Elements: Al, Cu, H_2; compounds: $Cu(NO_3)_2$, NaOH,

$Cu(OH)_2$, CuO, HCl, $CuCl_2$, $AlCl_3$

4. Analyzing Methods In the last step of the experiment, where is the aluminum chloride? How could you recover it?

Aluminum chloride is in the solution. Evaporate the water.

5. Inferring Conclusions What is the color of solutions of copper compounds?

blue or green

6. Analyzing Results Which substances involved in this experiment dissolve in water? Which do not dissolve?

Soluble: copper(II) nitrate, sodium hydroxide, hydrogen chloride,

copper(II) chloride, aluminum chloride

Insoluble: copper(II) hydroxide, copper(II) oxide, copper metal,

aluminum metal

7. Organizing Conclusions Refer to the procedure steps in the experiment, and complete the following word equations.
a. copper(II) nitrate + sodium hydroxide =

copper(II) hydroxide + sodium nitrate

b. copper(II) hydroxide + heat =

copper(II) oxide + water

c. copper(II) oxide + hydrochloric acid =

copper(II) chloride + water

d. copper(II) chloride + aluminum =

copper + aluminum chloride

e. hydrochloric acid + aluminum =

hydrogen + aluminum chloride

GENERAL CONCLUSIONS

1. Analyzing Information List four types of observations that indicate when a chemical change has occurred.

formation of a precipitate, color change, formation of a gas,

evolution of heat

2. **Relating Ideas** The chemical conversion of one product into another useful product is the basis for recycling. Explain how the type of reactions you observed in this experiment could be useful in the recycling of copper.

Copper metal was used in the preparation of the original

copper(II) nitrate solution. After several conversions, copper

metal was again recovered.

_____

_____

3. **Analyzing Conclusions** Describe the advantages and disadvantages of recycling metals as was done in this experiment.

Advantages: pure metal is obtained from a compound, the

reaction is done in one vessel.

Disadvantages: the reactions take time and may be

expensive, the waste products of recovery may cause pollution.

_____

_____

Name _____

Date _____ Class _____

Calcium and Its Compounds

OBJECTIVES

Recommended time:
60 min

- **Observe** some of the properties of calcium metal.
- **Determine** some of the chemical properties of calcium compounds.
- **Investigate** the properties of hard and soft water.

INTRODUCTION

**Solutions/
Materials
Preparation**

1. 15 pea-sized pieces of calcium

2. 50 g of CaO

3. 20 g of CaCl₂

4. To prepare the phenolphthalein solution, dissolve 1.00 g of phenolphthalein in 50 mL of denatured alcohol, and add 50 mL of distilled water.

Though calcium metal has a limited number of uses, its compounds are very common. The addition of calcium oxide to water produces a solution of calcium hydroxide. CaO is commonly called lime and the aqueous solution of $Ca(OH)_2$ is commonly called limewater. The formation of $Ca(OH)_2$ is a process called slaking, and solid $Ca(OH)_2$ is commonly called slaked lime.

When CO_2 is passed through limewater, it reacts to form $CaCO_3$. If more CO_2 is added to the system, it reacts to form $Ca(HCO_3)_2$. This reaction is reversible and can be represented by the following equation.

$$CaCO_3(s) + H_2O(l) + CO_2(g) \rightleftarrows Ca(HCO_3)_2(aq)$$

Calcium and magnesium compounds that are naturally present in tap water are responsible for a property known as "hardness." Hardness in water causes problems such a boiler scale, which is a residue that collects in pipes and can eventually clog them. Hard water also reduces the ability of soaps to form suds. It also produces soap scum in sinks and bathtubs. This scum is a precipitate formed from the reaction of the calcium ion with sodium stearate, which is an ingredient in some soaps.

SAFETY

5. The liquid soap must contain a sodium salt, such as Ivory® soap. Prepare about 500 mL of a concentrated solution of the soap for each class. Keep the solution concentration standard for each class.

6. The ion-exchange column is made from a large diameter tube filled with ion-exchange resin. The resin can be purchased from a chemical supply house.

 Always wear safety goggles and a lab apron to protect your eyes and clothing. If you get a chemical in your eyes, immediately flush the chemical out at the eyewash station while calling to your teacher. Know the location of the emergency lab shower and eyewash station and the procedure for using them.

 Do not touch any chemicals. If you get a chemical on your skin or clothing, wash the chemical off at the sink while calling to your teacher. Make sure you carefully read the labels and follow the precautions on all containers of chemicals that you use. If there are no precautions stated on the label, ask your teacher what precautions you should follow. Do not taste any chemicals or items used in the laboratory. Never return leftover chemicals to their original containers; take only small amounts to avoid wasting supplies.

Required Precautions

• Read all safety precautions and discuss them with your students. Students should be especially careful when boiling the solutions in the test tubes.

• Safety goggles and a lab apron must be worn at all times.

Call your teacher in the event of a spill. Spills should be cleaned up promptly, according to your teacher's directions.

When using a Bunsen burner, confine long hair and loose clothing. If your clothing catches on fire, WALK to the emergency lab shower, and use it to put out the fire. Do not heat glassware that is broken, chipped, or cracked. Use tongs or a hot mitt to handle heated glassware and other equipment because heated glassware and other equipment does not look hot. When heating liquid in a test tube, the mouth of the test tube should point away from where you or others are standing. Watch the test tube at all times to prevent the contents from boiling over.

MATERIALS

Disposal

All solutions can be poured down the drain. Solids may be put in the trash.

- 50 mL beaker
- 250 mL beaker
- Bunsen burner and related equipment
- calcium chloride
- calcium metal (small pieces)
- calcium oxide
- filter paper
- funnel
- glass tubing or plastic straw
- 50 mL graduated cylinder
- liquid soap
- matches
- medicine dropper
- phenolphthalein indicator
- ring and ring stand
- rubber stoppers to fit test tubes
- spatula
- 18 × 150 mm test tubes, 3
- test tube clamp
- test-tube rack
- wooden splint

Optional
- ion-exchange column

PROCEDURE

1. Observe the hardness, color, and general appearance of calcium metal. Record your observations.

Observations:

Calcium is a soft silvery-white metal. The metal pieces may show

some oxidation on the surface.

2. Break off a piece of calcium metal and place it in a beaker of water. Place a test tube over the metal in the beaker and collect the gas that is given off. When the test tube fills with gas, use your finger to stopper the test tube.

Remove the test tube from the beaker and test the gas with a lighted wood splint. Record your observations.

Observations:

Students should note that the gas does not support combustion.

---

3. Test the water in the beaker by adding 3 or 4 drops of phenolphthalein. Record your observations.

Observations:

The solution should turn pink indicating the presence of a base.

---

4. Place a spatula full of CaO in a test tube containing 10 mL of water. Touch the bottom of the test tube to detect any noticeable temperature change caused by this reaction. Record your observations.

Observation:

The test tube should feel warm.

Stopper and shake the test tube to be sure that all of the CaO has reacted. Pour off a few drops of the liquid into a separate test tube and add a few drops of phenolphthalein. Record your observations.

Observations:

The solution should turn pink indicating the presence of a base.

---

5. Filter the remaining solution and save the filtrate for Step **7.** The filtrate is limewater.

6. Obtain about 30 mL of liquid soap from your teacher. (Note this is a special mixture of liquid soap. Liquid detergent cannot be used). Pour 5 mL of distilled water into a clean test tube. Add liquid soap dropwise, using the medicine dropper, shaking vigorously after each drop. Count the drops and continue adding drops of liquid soap until the soapsuds last for 2 min. Record the number of drops used below. Also note the thickness of the suds.

Observations:

Number of drops of soap added   Student answers will vary depending

on the soap solution used.

Thickness of the suds   Student answers will vary.

7. Add 5 mL of the limewater filtrate from Step **5,** to about 20 mL of distilled water in a small beaker. Using a piece of glass tubing or a straw, blow into this solution until it becomes cloudy. Continue blowing into the solution until it becomes clear. The resulting solution contains $Ca(HCO_3)_2$. This solution can be considered "temporary" hard water. Pour 5 mL of this solution into each of two text tubes labeled A and B. Boil the solution in test tube A for a few minutes and record any changes. Allow the contents of test tube A to cool.

Observations:

Students should notice the formation of a white precipitate.

8. Add the liquid soap dropwise to the cooled solution in test tube A. Shake vigorously after each drop and record the number of drops added to cause the suds to remain for two minutes.

Observations:

Number of drops of soap added   Student answers will vary.

9. Repeat step 8 with the solution in test tube B. Record the number of drops of soap needed to produce the same level of suds as in test tube A.

Observations:

Number of drops of soap added   Student should notice much more

soap is needed than in Step 8.

10. Add about 0.5 g of $CaCl_2$ to 50 mL of distilled water in a clean beaker. Stir this solution until the $CaCl_2$ is dissolved. This solution is "permanent" hard water. Pour 5 mL of this solution into each of three test tubes labeled A, B, and C. Boil the sample in test tube A for a few minutes and test the sample with liquid soap for hardness. Record the number of drops of soap added to cause the suds to remain for two minutes.

Observations:

Number of drops of soap added   Student answers will vary.

11. Add soap to the sample in test tube B. Record the number of drops of soap added to cause the suds to remain for two minutes.

Observations:

Number of drops of soap added   Student answers will vary, but they

should notice no difference between the number of drops needed

in this step vs. Step 10.

12. **Optional:** Pour the sample in test tube C through an ion-exchange column set up by your teacher. Collect the filtrate in a test tube and add soap to the sample. Record the number of drops of soap added to cause the suds to remain for two minutes.

Observations:

Number of drops of soap added   Student answers will vary, but they

should notice that fewer drops of soap are used than in Steps 10

and 11 because the water has been softened.

Cleanup and Disposal

13. Clean all apparatus and your lab station. Return equipment to its proper place. Dispose of chemicals and solutions in the containers designated by your teacher. Do not pour any chemicals down the drain or into the trash unless your teacher directs you to do so. Wash your hands thoroughly before you leave the lab and after all work is finished.

QUESTIONS

1. **Analyzing Results** Write an equation for the reaction of calcium and water. How do you know what gas was produced?

   $Ca(s) + 2H_2O(l) \rightarrow Ca(OH)_2(aq) + H_2(g)$

   Because the gas did not support combustion it has to be

   hydrogen.

2. Write an equation for the slaking of lime.

   $CaO(s) + H_2O(l) \rightarrow Ca(OH)_2(aq)$

3. List the solute present in the "temporary" hard water.

   The temporary hard water contains $Ca(HCO_3)_2$.

4. Explain, using an equation, how boiling "softens" the temporary hard water. Hint: Look at the reversible reaction for this process given in the Introduction. Cooling favors the forward reaction. Heating favors the reverse reaction.

   The temporary hard water contains a calcium hydrogen carbonate

   compound. Hydrogen carbonates are thermally unstable and de-

   compose upon heating.

   $Ca(HCO_3)_2(aq) \xrightarrow{heat} CaCO_3(s) + H_2O(l) + CO_2(g)$

5. **Making Predictions** Use your answer from item 4 to make a prediction about why permanent hard water is not affected by the boiling process.

   The compounds in the permanent hard water must be thermally

   stable and are not affected by the boiling process. Therefore, they

   do not break down upon heating to soften the water.

6. **Making Predictions** How do you think an ion-exchange column softens water? Hint: Look back at the Introduction to recall what types of compounds are in hard water.

Ion-exchange columns remove Ca^{2+} ions by exchanging them

with Na^+ ions that will not form insoluble salts with compounds

in the water.

7. **Inferring Conclusions** What is one advantage of installing a water-softening unit if you live in an area that has hard water?

You need less soap to produce good sudsing action when water is

soft.

8. **Inferring Conclusions** Sodium chloride is a major component of most water-softening units. Make a prediction as to how a water-softening unit works.

The sodium chloride in the water-softening unit provides a source

of Na^+ ions to exchange with Ca^{2+} and Mg^{2+} ions. The sodium

compounds formed are water soluble, therefore the water is now

soft.

ChemFile

EXPERIMENT **A7**

Water of Hydration

OBJECTIVES

Recommended time:
60 min

- **Determine** that all the water has been driven from a hydrate by heating a sample to constant mass.

- **Use** experimental data to calculate the moles of water released by a hydrate.

- **Infer** the empirical formula of the hydrate from the formula of the anhydrous compound and experimental data.

INTRODUCTION

Many ionic compounds, when crystallized from an aqueous solution, will take up definite amounts of water as an integral part of their crystal structures. This water of crystallization may be driven off by heating the hydrated substance to convert it to its anhydrous form. Because the law of definite composition holds for crystalline hydrates, the number of moles of water driven off per mole of the anhydrous compound is a simple whole number. If the formula of the anhydrous compound is known, you can use your data to determine the formula of the hydrate.

SAFETY

Solution/Material Preparation

1. If you don't have a desiccator, you can use clean, empty jars with wide mouths and tight lids, such as peanut butter jars. Put a layer of granular anhydrous calcium chloride on the bottom of the jar. Bend down the ends of pipestem triangles to act as legs. Place one triangle in each jar to hold a crucible securely.

Required Precautions

- Safety goggles and a lab apron must be worn at all times.

- Read all safety precautions, and discuss them with your students.

  Always wear safety goggles and a lab apron to protect your eyes and clothing. If you get a chemical in your eyes, immediately flush the chemical out at the eyewash station while calling to your teacher. Know the location of the emergency lab shower and the eyewash station and the procedure for using them.

 Do not touch any chemicals. If you get a chemical on your skin or clothing, wash the chemical off at the sink while calling to your teacher. Make sure you carefully read the labels and follow the precautions on all containers of chemicals that you use. If there are no precautions stated on the label, ask your teacher what precautions you should follow. Do not taste any chemicals or items used in the laboratory. Never return leftovers to their original containers; take only small amounts to avoid wasting supplies.

 Call your teacher in the event of a spill. Spills should be cleaned up promptly, according to your teacher's directions.

 When using a Bunsen burner, confine long hair and loose clothing. If your clothing catches on fire, WALK to the emergency lab shower and use it to put out the fire. Do not heat glassware that is broken, chipped, or cracked. Use tongs or a hot mitt to handle heated glassware and other equipment because hot glassware does not always look hot.

• Remind students to confine loose clothing and long hair before lighting a burner.

 Never put broken glass in a regular waste container. Broken glass should be disposed of properly.

MATERIALS

• Remind students that heated objects can be hot enough to burn even if they look cool. Students should always use crucible tongs to handle crucibles and covers.

• balance, centigram
• Bunsen burner and related equipment
• crucible and cover
• desiccator
• iron ring
• magnesium sulfate, Epsom salts, hydrated crystals
• pipe-stem triangle
• ring stand
• sparker
• spatula
• tongs

PROCEDURE

Techniques to Demonstrate

Set up the ring stand, ring, and pipe-stem triangle with the crucible in place. Show students how to place the crucible cover slightly ajar. Turn on and adjust the burner flame to the small, hot flame that is needed to dehydrate the salt. Demonstrate how to use tongs to transfer the hot crucible to the desiccator.

Pre-lab Discussion

Caution students that to avoid loss of some of the hydrate, the opening left by the crucible cover must not be too large. Also, the crucible should not be heated too strongly for the first few minutes because some of the hydrate could be lost with the water vapor.

1. Throughout the experiment, handle the crucible and cover with clean crucible tongs only. Place the crucible and cover on the triangle as shown in Figure A. Position the cover slightly tipped, leaving only a small opening for any gases to escape. Preheat the crucible and cover to redness.
CAUTION The crucible and cover are very hot after each heating. Remember to handle them only with tongs.

2. Using tongs, transfer the crucible and cover to a desiccator. Allow them to cool 5 min in the desiccator. Never place a hot crucible on a balance. When cool, determine the mass of the crucible and cover to the nearest 0.01 g. Record this mass in your data table.

3. Using a spatula, add approximately 5 g of magnesium sulfate hydrate crystals to the crucible. Determine the mass of the covered crucible and crystals to the nearest 0.01 g. Record this mass in your data table.

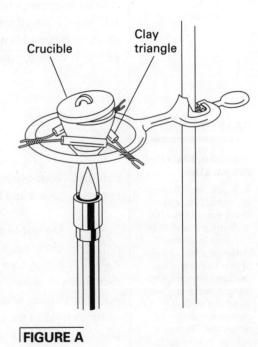
Crucible   Clay triangle

FIGURE A

4. Place the crucible with the magnesium sulfate hydrate on the triangle, and again position the cover so that there is a small opening. Too large an opening may allow the hydrate to spatter out of the crucible. Heat the crucible very gently with a low flame to avoid spattering any of the hydrate. Increase the temperature gradually for 2 or 3 min. Then, heat strongly, but not red-hot, for at least 5 min.

Disposal

Provide a labeled container for students to dispose of the rehydrated and anhydrous magnesium sulfate and any excess of the original compound. Later, redissolve the contents of the container in distilled water. Let the solutions evaporate until dry, and then recover the crystals for reuse next year. If students repeat the experiment using any of the hydrates listed in General Conclusions **3,** dispose of the Na_2CO_3, Na_2SO_4, $NaAl(SO_4)_2$, KCl, and $MgCl_2$ by dissolving the salts in water and pouring them down the drain. Dissolve $CuSO_4$ in water, add sufficient 1.0 M NaOH to precipitate the copper, and then filter. The pH of the filtrate should be adjusted to between 5 and 9 and then poured down the drain. The precipitate can be dried and put in the trash.

5. Using tongs, transfer the crucible, cover, and contents to the desiccator, and allow them to cool for 5 min. Then, using the same balance you used in step **2,** determine their mass. Be sure the crucible is sufficiently cool because heat can affect your measurement. Record the mass in your data table.

6. Again heat the covered crucible and contents strongly for 5 min. Allow the crucible, cover, and contents to cool in the desiccator, and then use the same balance as before to determine their mass. If the last two mass measurements differ by no more than 0.01 g, you may assume that all the water has been driven off. Otherwise, repeat the heating process until the mass no longer changes. Record this constant mass in your data table.

Cleanup and Disposal

7. Clean all apparatus and your lab station. Return equipment to its proper place. Dispose of the $MgSO_4$ in your crucible as your teacher directs. Do not pour any chemicals down the drain or in the trash unless your teacher directs you to do so. Wash your hands thoroughly after all work is finished and before you leave the lab.

Data Table

Mass of empty crucible and cover	32.25 g
Mass of crucible, cover, and magnesium sulfate hydrate	37.25 g
Mass of crucible, cover, and anhydrous magnesium sulfate after 1st heating	34.73 g
Mass of crucible, cover, and anhydrous magnesium sulfate after 2nd heating	34.72 g

CALCULATIONS

1. Organizing Data Calculate the mass of anhydrous magnesium sulfate (the residue that remained after driving off the water).

34.72 g − 32.25 g = 2.47 g anhydrous $MgSO_4$ (mass after 2nd heating)

2. Organizing Data Calculate the moles of anhydrous magnesium sulfate.

$$2.47 \text{ g} \times \frac{1 \text{ mol MgSO}_4}{120.37 \text{ g}} = 0.0205 \text{ mol anhydrous MgSO}_4$$

3. Organizing Data Calculate the mass of water driven off from the hydrate.

37.25 g − 34.72 g = 2.53 g H_2O

4. Organizing Data Calculate the number of moles of water driven off from the hydrate.

$$2.53 \text{ g} \times \frac{1 \text{ mol H}_2\text{O}}{18.02 \text{ g}} = 0.141 \text{ mol H}_2\text{O}$$

5. **Organizing Ideas** Write the equation for the reaction that occurred when you heated hydrated $MgSO_4$ in this experiment. Use the letter n to represent the number of moles of water driven off per mole of anhydrous magnesium sulfate.

$MgSO_4 \cdot nH_2O(s) \rightarrow MgSO_4(s) + nH_2O(g)$

6. **Organizing Conclusions** Using your answers to Calculations items **2**, **4**, and **5**, determine the mole ratio of $MgSO_4$ to H_2O to the *nearest whole number.*

$$\text{Mole ratio} = \frac{H_2O}{MgSO_4} = \frac{0.141 \text{ mol}}{0.0205 \text{ mol}} = \frac{6.88 \text{ mol } H_2O}{1 \text{ mol } MgSO_4}; 1:7$$

7. **Organizing Conclusions** Use your answer to Calculations item **6** to write the formula for the magnesium sulfate hydrate.

$MgSO_4 \cdot 7H_2O$

Calculations Table

Mass of anhydrous magnesium sulfate	2.47 g
Moles of anhydrous magnesium sulfate	0.0205 mol
Mass of water driven off from hydrate	2.53 g
Moles of water driven off from hydrate	0.141 mol
Mole ratio of water to anhydrous magnesium sulfate	6.88 : 1.00
Empirical formula of the hydrate	$MgSO_4 \cdot 7H_2O$

QUESTIONS

1. How does this experiment exemplify the law of definite composition?

   Every compound has the same elements in exactly the same

   proportion by mass no matter what the size of the sample or its

   source. You can infer that there will always be the same mass, and

   the same number of moles, of water molecules present in each

   mole of magnesium sulfate heptahydrate.

**GENERAL
CONCLUSIONS**

1. **Applying Conclusions** The following results were obtained when a solid was heated by three different lab groups. In each case, the students observed that when they began to heat the solid, drops of a liquid formed on the sides of the test tube.

Lab group	Mass before heating	Mass after heating
1	1.48 g	1.26 g
2	1.64 g	1.40 g
3	2.08 g	1.78 g

a. Could the solid be a hydrate? What evidence supports your answer?

Yes, because after heating, the mass is less, indicating that

some mass—probably H_2O—was driven off.

b. If, after heating, the solid has a molar mass of 208 g/mol and a formula of XY, what is the formula of the hydrate?

Moles of anhydrous salt(XY) $= 1.26 \text{ g} \times \dfrac{1 \text{ mol}}{208 \text{ g}} = 0.00606 \text{ mol}$

mol $H_2O = 0.22 \text{ g} \times \dfrac{1 \text{ mol}}{18.02 \text{ g}} = 0.012 \text{ mol}$

$\dfrac{\text{mol } H_2O}{\text{mol XY}} = \dfrac{0.012}{0.0061} = \dfrac{2}{1}$; the formula is XY $\cdot 2H_2O$

2. **Applying Conclusions** Some cracker tins include a glass vial of drying material in the lid to keep the crackers crisp. The material is often a mixture of magnesium sulfate and cobalt chloride indicator. As the magnesium sulfate absorbs moisture ($MgSO_4 \cdot H_2O + 6H_2O \rightarrow MgSO_4 \cdot 7H_2O$), the indicator changes color from blue to pink ($CoCl_2 \cdot 4H_2O + 2H_2O \rightarrow CoCl_2 \cdot 6H_2O$). When this drying mixture becomes totally pink, it can be restored by heating in an oven. What two changes are caused by the heating?

Heating removes some of the water of hydration from both the

magnesium sulfate drying agent and the cobalt chloride indicator.

The mixture turns back to a blue color, ready to absorb more

moisture.

3. **Analyzing Methods** Why did you use the same balance each time you measured the mass of the crucible and its contents?

There can be differences in balances. For the most accurate results

when comparing masses, it's best to use the same balance each

time.

4. **Predicting Outcomes** Use handbooks to investigate the properties of the compounds listed in the following table. Write answers to these questions in the table: Does the compound form a hydrate? How would you describe the anhydrous compound? If your teacher approves, repeat the experimental procedure using one of the listed compounds and verify its hydrate formula. Explain any large deviation from the correct hydrate formula.

Name of compound	Water of crystalliza-tion?	Description of anhydrous compound
Sodium carbonate	yes	white powder
Sodium sulfate	yes	white powder
Sodium aluminum sulfate	yes	white powder
Potassium chloride	no	unchanged
Magnesium chloride	yes	white powder
Copper sulfate	yes	white powder

Name _____

Date _____ Class _____

EXPERIMENT **A8**

Boyle's Law

OBJECTIVES

Recommended
time: 30 minutes

- **Determine** the volume of a gas in a container under various pressures.
- **Graph** pressure-volume data to discover how the variables are related.
- **Interpret** graphs and verify Boyle's law.

INTRODUCTION

**Solution/Materials
Preparation**

1. Check each apparatus for possible leaks. If the apparatus leaks, the piston will not return to its original position regardless of the number of times you twist the head. The most probable source of leakage is the bottom end of the syringe where the needle is normally attached.

According to Boyle's law, the volume of a fixed amount of dry gas is inversely proportional to the pressure, if the temperature is constant. Boyle's law may be stated mathematically as $P \propto 1/V$ or $PV = k$ (where k is a constant). Notice that for Boyle's law to apply, two variables that affect gas behavior must be held constant: the amount of gas and the temperature.

In this experiment, you will vary the pressure of air contained in a syringe and measure the corresponding change in volume. Because it is often impossible to determine relationships by just looking at data in a table, you will plot graphs of your results to see how the variables are related. You will make two graphs, one of volume versus pressure and another of the inverse of volume versus pressure. From your graphs you can derive the mathematical relationship between pressure and volume and verify Boyle's law.

SAFETY

  Always wear safety goggles and a lab apron to protect your eyes and clothing. If you get a chemical in your eyes, immediately flush the chemical out at the eyewash station while calling to your teacher. Know the location of the emergency lab shower and the eyewash station and the procedure for using them.

MATERIALS

- Boyle's law apparatus
- carpet thread
- objects of equal mass (approximately 500 g each), 4

PROCEDURE

1. Adjust the piston head of the Boyle's law apparatus so that it reads between 30 and 35 cm³. To adjust, pull the piston head all the way out of the syringe, insert a piece of carpet thread in the barrel, and position the piston head at the desired location, as shown in Figure A on the next page.
Note: Depending upon the Boyle's law apparatus that you use, you may find the volume scale on the syringe abbreviated in cc or cm³. Both abbreviations stand for cubic centimeter (1 cubic centimeter is equal to 1 mL). The apparatus shown in Figure A is marked in cm³.

2. Each weight should be approximately 500 g. Less mass provides too small a change in volume. Greater mass may cause the pressure to be so excessive that air will begin to escape around the gasket. 500 g weights are ideal. All weights used by one lab group should be the same.

Required Precautions

• Read all safety precautions, and discuss them with your students.
• Safety goggles and a lab apron must be worn at all times.

Procedural Tips

• Show students how to give the piston several twists each time they change the pressure. The frictional force is significant, but if the piston is twisted until a stable volume is reached, reliable data should be obtained. Show students how to carefully stack the weights so that the pile is stable.
• You may wish to have students calculate the uncertainty of the volumes and plot these on the graph of volume versus pressure. This could lead to a discussion about uncertainty and experimental error. In addition, it may help students appreciate why a line which does not actually go through every point is nevertheless an accurate representation of the data.

Disposal

None

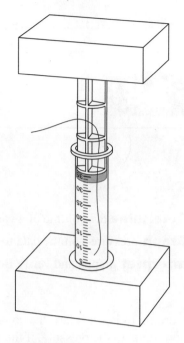

FIGURE A

2. While holding the piston in place, carefully remove the thread. Twist the piston several times to allow the head to overcome any frictional forces. Read the volume to the nearest 0.1 cm³. Record this value in your data table as the initial volume for zero weights.

3. Place one of the weights on the piston. Give the piston several twists to overcome any frictional forces. When the piston comes to rest, read and record the volume to the nearest 0.1 cm³.

4. Repeat step **3** for two, three, and four weights, and record your results.

5. Repeat steps **3** and **4** for at least two more trials. Record your results.

Data Table			
Pressure (number of weights)	**Trial 1 Volume (cm³)**	**Trial 2 Volume (cm³)**	**Trial 3 Volume (cm³)**
0	33.0	33.0	33.0
1	29.7	29.5	29.0
2	26.8	26.8	26.0
3	24.3	24.1	23.5
4	22.0	21.8	21.2

Cleanup and Disposal

6. Clean all apparatus and your lab station at the end of this experiment. Return equipment to its proper place. Ask your teacher how to dispose of all waste materials. Wash your hands thoroughly before leaving the lab and after all work is finished.

CALCULATIONS

1. **Organizing Data** Calculate the average volume of the three trials for weights 0–4. Record your results in your calculations table.

$$0 \text{ weight} = \frac{33.0 + 33.0 + 33.0}{3} = 33.0 \text{ cm}^3$$

$$1 \text{ weight} = \frac{29.7 + 29.5 + 29.0}{3} = 29.4 \text{ cm}^3$$

$$2 \text{ weights} = \frac{26.8 + 26.8 + 26.0}{3} = 26.5 \text{ cm}^3$$

$$3 \text{ weights} = \frac{24.3 + 24.1 + 23.5}{3} = 24.0 \text{ cm}^3$$

$$4 \text{ weights} = \frac{22.0 + 21.8 - 21.2}{3} = 21.7 \text{ cm}^3$$

2. **Organizing Data** Calculate the inverse for each of the average volumes. For example, if the average volume for three weights is 26.5 cm^3, then $1/V = 1/26.5/\text{cm}^3 = 0.0377/\text{cm}^3$.

$$0 \text{ weight} = \frac{1}{33.0} - 3.03 \times 10^{-2}/\text{cm}^3$$

$$1 \text{ weight} = \frac{1}{29.4} - 3.40 \times 10^{-2}/\text{cm}^3$$

$$2 \text{ weights} = \frac{1}{26.5} - 3.77 \times 10^{-2}/\text{cm}^3$$

$$3 \text{ weights} = \frac{1}{24.0} - 4.17 \times 10^{-2}/\text{cm}^3$$

$$4 \text{ weights} = \frac{1}{21.7} - 4.61 \times 10^{-2}/\text{cm}^3$$

Calculations Table

Pressure (number of weights)	Average Volume (cm^3)	1/Volume ($\times 10^2/\text{cm}^3$)
0	33.0	3.03
1	29.4	3.40
2	26.5	3.77
3	24.0	4.17
4	21.7	4.61

QUESTIONS

1. **Analyzing Data** Plot a graph of volume versus pressure on the grid below. Because the number of weights added to the piston is directly proportional to the pressure applied to the gas, you can use the number of weights to
represent the pressure. Notice that pressure is plotted on the horizontal axis
and volume is plotted on the vertical axis. Draw the smoothest curve that
goes through most of the points.

Volume Versus Pressure

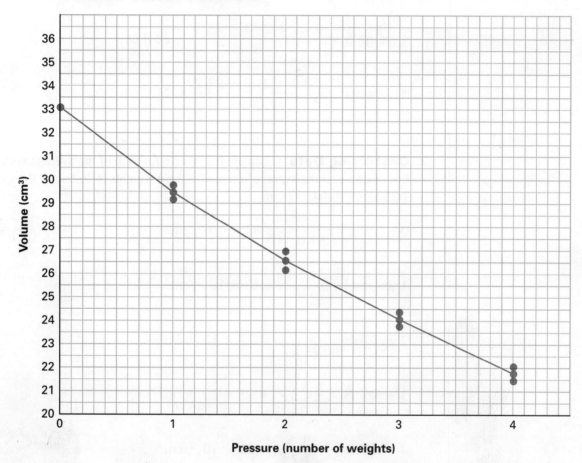

2. **Interpreting Graphics** Does your graph indicate that a change in
volume is directly proportional to a change in pressure? Explain your
answer.

   No, the graph is not a straight line. Furthermore, as the pressure

   increases, the volume decreases.

ChemFile

3. Analyzing Data On the grid below, plot a graph of 1/volume versus pressure. Notice that pressure is on the horizontal axis and 1/volume is on the vertical axis. Draw the best line that goes through the majority of the points.

Versus 1/V Pressure

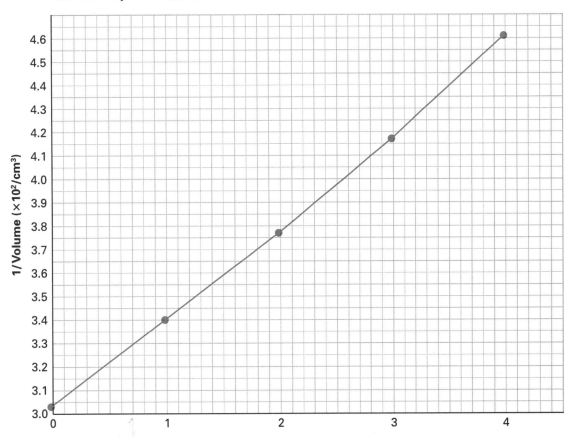

4. Interpreting Graphics What do you conclude about the mathematical relationship between the pressure applied to a gas and its corresponding volume?

The graph of 1/volume versus pressure is approximately a straight

line (within experimental uncertainty). Therefore, pressure must

be proportional to 1/volume. Pressure and volume are inversely

related.

_____

_____

GENERAL CONCLUSIONS

1. **Applying Models** Correlate the observed relationship between the pressure and volume of a gas with the kinetic theory description of a gas.

Gas pressure results from particles hitting the sides of the container. If the size of the container is decreased while the number of molecules remains the same, the number of hits is increased.

2. **Applying Conclusions** Use your graph to predict the volume of the gas if 2.5 weights were used.

Answers will vary

3. **Applying Conclusions** If a normal sea-level recipe is used to prepare a cake at a location 1000 meters below the surface of Earth, the cake will be much flatter than expected. Explain why, and offer a solution. (Hint: Consider how barometric pressure differs at this altitude and what effect that might have on the ability of the cake to rise.)

At the lower altitude, the barometric pressure is greater. As a result, Boyle's law predicts that the gas bubbles will be smaller. To compensate for the reduced volume of gas, more baking powder or less flour can be added to the batter.

Name _____

Date _____ Class _____

EXPERIMENT

Molar Volume of a Gas

OBJECTIVES

Recommended time:
60 min

- **Determine** the volume of hydrogen produced by the reaction of a known mass of magnesium with hydrochloric acid.
- **Compute** the volume of the gas at standard temperature and pressure.
- **Relate** the volume of gas to the moles of magnesium reacted.
- **Infer** the volume of one mole of gas at standard temperature and pressure.

INTRODUCTION

Magnesium is an active metal that readily reacts with hydrochloric acid to produce hydrogen gas. In this experiment, you will react a known amount of magnesium with hydrochloric acid and collect and measure the volume of hydrogen gas produced. From the volume of the gas measured at atmospheric pressure and the temperature of the lab, you can calculate the volume of the gas at standard temperature and pressure. Knowing the mass of magnesium, you can calculate the moles of magnesium consumed and determine the volume of hydrogen at STP produced by the reaction of one mole of magnesium. This conclusion can then be related to the balanced equation for the reaction.

SAFETY

Solution/ Material Preparation

1. To prepare 1 L of 6 M hydrochloric acid, observe the required safety precautions. Slowly and with stirring, add 516 mL of concentrated HCl to enough distilled water to make 1.0 L of solution.

Required Precautions

- Read all safety precautions and discuss them with your students.

- Safety goggles and a lab apron must be worn at all times.

  Always wear safety goggles and a lab apron to protect your eyes and clothing. If you get a chemical in your eyes, immediately flush the chemical out at the eyewash station while calling to your teacher. Know the location of the emergency lab shower and the eyewash station and the procedure for using them.

 Do not touch any chemicals. If you get a chemical on your skin or clothing, wash the chemical off at the sink while calling to your teacher. Make sure you carefully read the labels and follow the precautions on all containers of chemicals that you use. If there are no precautions stated on the label, ask your teacher what precautions you should follow. Do not taste any chemicals or items used in the laboratory. Never return leftovers to their original containers; take only small amounts to avoid wasting supplies.

 Call your teacher in the event of a spill. Spills should be cleaned up promptly, according to your teacher's directions.

 Never put broken glass in a regular waste container. Broken glass should be disposed of properly according to your teacher's instructions.

MATERIALS

- In case of a spill, use a dampened cloth or paper towel (or more than one towel if necessary) to mop up the spill. Then rinse the towel in running water at the sink, wring it out until it is only damp, and put it in the trash. In the event of a spill of an acid, first dilute the spill with water and then proceed as described.

- 6 M HCl
- 50 mL beaker
- 50 mL eudiometer
- 400 mL beaker
- 1000 mL graduated cylinder or hydrometer jar
- buret clamp
- magnesium ribbon (untarnished)
- ring stand
- rubber stopper, one-hole, #00
- thermometer, nonmercury, 0–100°C
- thread

PROCEDURE

- Wear safety goggles, a face shield, impermeable gloves, and a lab apron when you prepare the HCl. Work in a hood known to be in operating condition and have another person stand by to call for help in case of an emergency. Be sure you are within a 30 second walk from a safety shower and eyewash station known to be in operating condition.

- Procedure step 5 should be done by the teacher for each lab group. Students should not work with 6 M HCl. Wear safety goggles, face shield, lab apron, and gloves while dispensing the HCl into the student eudiometers. Students should stand clear and wear safety goggles and lab aprons.

1. Fill a 400 mL beaker two-thirds full of water that has been adjusted to room temperature.

2. Measure a length of magnesium ribbon to the nearest 0.1 cm. Your piece of magnesium should not exceed 4.5 cm. Record the length of the ribbon in the Data Table.

3. Obtain the mass of one meter of magnesium ribbon from your teacher and record this mass.

4. Roll the length of the magnesium ribbon into a loose coil. Tie it with one end of a piece of thread approximately 25 cm in length. All the loops of the coil should be tied together as shown in Figure A.

5. This step requires the use of 6 M hydrochloric acid and will be performed by your teacher.
CAUTION Hydrochloric acid is caustic and corrosive. Avoid contact with skin and eyes. Avoid breathing the vapor. Make certain that you are wearing safety goggles, a lab apron, and gloves when working with the acid. If any acid should splash on you, immediately flush the area with water and then report the incident to your teacher. If you should spill any acid on the counter top or floor, ask your teacher for the appropriate spill package to be used in the cleanup.
Your teacher will carefully pour approximately 10 mL of 6 M HCl into a 50 mL beaker and then pour the 10 mL of HCl into your gas measuring tube or eudiometer.

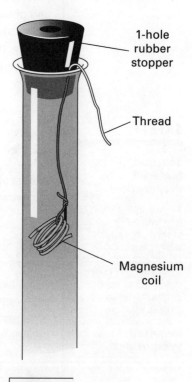

1-hole rubber stopper

Thread

Magnesium coil

FIGURE A

6. While holding the eudiometer in a slightly tipped position, very slowly pour water from the 400 mL beaker into the eudiometer, being careful to layer the

ChemFile

water over the acid so that they do not mix. Add enough water to fill the eudiometer to the brim.

7. Lower the magnesium coil into the water in the eudiometer tube to a depth of about 5 cm. Insert the rubber stopper into the open end of the eudiometer to hold the thread in place, as shown in Figure A on the previous page. The one-hole stopper should displace some water from the tube. This ensures that no air is left inside the tube.

8. Cover the hole of the stopper with your finger, and invert the eudiometer into the 400 mL beaker of water. Clamp the eudiometer tube into position on the ring stand, as shown in Figure B. The acid flows down the tube (why?) and reacts with the magnesium. Is the acid now more concentrated or dilute? Answers these questions and describe your observations in the place provided following the Data Table.

9. When the magnesium has disappeared entirely and the reaction has stopped, cover the stopper hole with a finger and carefully transfer the eudiometer tube to a 1000 mL graduated cylinder or other tall vessel that has been filled with water. Adjust the level of the eudiometer tube in the water as shown in Figure C. The levels of the liquid inside the eudiometer and the cylinder should be the same. Read the volume of hydrogen you collected as accurately as possible.

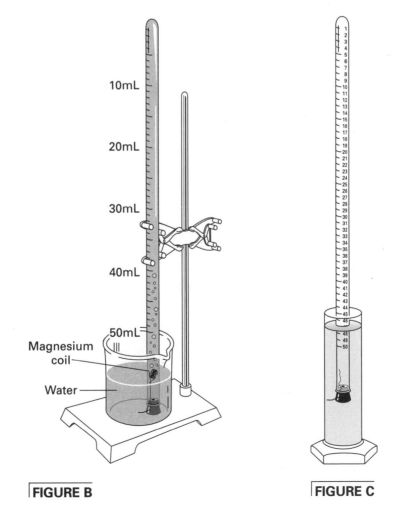

| **FIGURE B** | **FIGURE C** |

10. Record the temperature of the room and the atmospheric pressure.

11. Use the table of water-vapor pressures in your textbook to find the vapor pressure of water at the temperature of the room. Record this water-vapor pressure in your data table.

Cleanup and Disposal

12. Clean all apparatus and your lab station. Return equipment to its proper place. Dispose of chemicals and solutions in the containers designated by your teacher. Do not pour any chemicals down the drain or in the trash unless your teacher directs you to do so. Wash your hands thoroughly before you leave the lab and after all work is finished.

Data Table

Length of Mg used	4.2	cm
Mass per meter of Mg	0.981	g/m
Volume of H_2 collected under lab conditions	42.6	mL
Temperature of H_2 collected	20.0	°C
Barometer reading	755	mm Hg
Vapor pressure of water at observed temperature	17.5	mm Hg

Observations:

The acid is denser than water. The acid is being diluted. Gas

bubbles begin to form and rise to the top of the liquid. The

magnesium ribbon starts to disappear.

CALCULATIONS

1. Organizing Data From the length of the magnesium ribbon you used and the mass of a meter of magnesium ribbon, determine the mass of the magnesium that reacted. Record this result and all your calculated results in your calculations table.

$$\text{Mass of Mg} = 4.2 \text{ cm} \times \frac{1 \text{ m}}{100 \text{ cm}} \times 0.981 \, \frac{\text{g}}{\text{m}} = 0.041 \text{ g}$$

2. Organizing Data Calculate the number of moles of magnesium that reacted.

$$\text{Moles of Mg} = 0.041 \text{ g} \times \frac{1 \text{ mol Mg}}{24.3 \text{ g}} \times 0.0017 \text{ mol}$$

3. **Organizing Data** Because the hydrogen gas was collected over water, two gases were present in the eudiometer: hydrogen and water vapor. Calculate the partial pressure of the hydrogen gas you collected at 20.0°C. (Hint: The sum of the two partial pressures equals atmospheric pressure.)

Pressure H_2 = 755 mm Hg − 17.5 mm Hg = 738 mm Hg

4. **Organizing Data** Calculate the volume of the dry hydrogen you collected at 20.0°C and standard pressure. (Hint: $P_1V_1 = P_2V_2$)

$$V_2 = \frac{P_1V_1}{P_2} = \frac{(738 \text{ mm Hg} \times 42.6 \text{ mL})}{760 \text{ mm Hg}} = 41.4 \text{ mL}$$

5. **Organizing Data** Calculate the volume of dry hydrogen gas at standard temperature. $\left(\text{Hint: } \dfrac{V_1}{T_1} = \dfrac{V_2}{T_2}\right)$

$$V_2 = \frac{V_1T_2}{T_2} = \frac{(41.4 \text{ mL} \times 273 \text{ K})}{293 \text{ K}} = 38.6 \text{ mL}$$

6. **Inferring Conclusions** Calculate the volume of dry hydrogen gas that would be produced by one mole of magnesium at standard temperature and pressure.

$$\frac{\text{Volume } H_2}{1 \text{ mol Mg}} = \frac{38.6 \text{ mL}}{0.0017 \text{ mol Mg}} = 23\ 000 \text{ mL/ mol}$$

1. **Analyzing Methods** Why was it necessary to make a water-vapor pressure correction of the barometer reading in this experiment?

When a gas is collected over water, some of the water vaporizes

and mixes with the collected gas. To obtain the true pressure of

the dry hydrogen, the water-vapor pressure must be subtracted

from the total pressure.

2. Analyzing Methods Why was it necessary to adjust the level of the eudiometer in the cylinder, as in Figure C, so that the level of water in the eudiometer was the same as the level of water in the cylinder?

When the water levels inside and outside the eudiometer are

equal, the pressure of the gas inside the tube is equal to atmos-

pheric pressure. If the level of water in the eudiometer was below

the level in the cylinder, the pressure of the gas would be greater

than atmospheric pressure. If the level in the eudiometer was

above the level in the cylinder, the pressure of the gas would be

less than atmospheric pressure.

GENERAL CONCLUSIONS

1. Relating Ideas Write a balanced equation for the reaction of magnesium with HCl. The products are hydrogen gas and $MgCl_2$.

$Mg(s) + 2HCl(aq) \rightarrow MgCl_2(aq) + H_2(g)$

2. Applying Conclusions From the balanced equation above, determine the volume of hydrogen gas at standard temperature and pressure that can be produced from three moles of magnesium reacting with hydrochloric acid.

Because there is a one-to-one mole ratio of Mg to H_2, 3 mol of

magnesium produce 3 mol of hydrogen gas. According to this ex-

periment, at STP this is a volume of 3 × 23 000 mL, or 69 000 mL.

3. Applying Conclusions Since the 1930s, alloys of aluminum and magnesium have been used in the manufacture of pots and pans. An unanswered question is whether a small amount of these metals that might dissolve in food as it cooks is beneficial, harmful, or of no consequence. What are some examples of foods that will react with aluminum/magnesium pans?

Acidic foods such as citrus fruits, rhubarb, sauerkraut, and pickles

will react with magnesium and aluminum.

Name _____

Date _____ Class _____

Molar Heat of Fusion of Ice

OBJECTIVES

Recommended time:
30 minutes

- **Measure** the temperature change of a mass of water as ice melts.
- **Calculate** the heat energy associated with the change in the temperature of the water.
- **Determine** the heat energy required to melt a mole of ice.

INTRODUCTION

Solution/Materials Preparation

None

Required Precautions

- Read all safety precautions, and discuss them with your students.

Water freezes and melts at the same temperature: 0°C. The heat energy required to melt one mole of ice at its melting point is its molar heat of fusion. The same amount of energy is released into the surroundings when a mole of liquid water freezes. For any substance, the molar heat of fusion is the heat required to melt one mole of that substance. The usual units are kilojoules per mole.

In this experiment, you will melt some ice in a foam cup containing warm water and measure the change in temperature of the water. Knowing the initial mass of the ice and the initial mass of the water, you will be able to calculate the molar heat of fusion of ice.

SAFETY

- Safety goggles and a lab apron must be worn at all times.
- In case of a spill, use a dampened cloth or paper towels to mop up the spill. Then rinse the cloth in running water at the sink, wring it out until it is only damp, and put it in the trash.

Procedural Tip

- Emphasize that students must use a stirring rod—and not the thermometer—for stirring the ice and water mixture in Procedure step **3**.

Disposal

None

  Always wear safety goggles and a lab apron to protect your eyes and clothing. If you get a chemical in your eyes, immediately flush the chemical out at the eyewash station while calling to your teacher. Know the location of the emergency lab shower and the eyewash station and the procedure for using them.

 Do not touch any chemicals. If you get a chemical on your skin or clothing, wash the chemical off at the sink while calling to your teacher. Make sure you carefully read the labels and follow the precautions on all containers of chemicals that you use. If there are no precautions stated on the label, ask your teacher what precautions you should follow. Never return leftovers to their original containers; take only small amounts to avoid wasting supplies.

 Call your teacher in the event of a spill. Spills should be cleaned up promptly, according to your teacher's directions.

 Never put broken glass in a regular waste container. Broken glass should be disposed of properly.

 CAUTION Before using the Bunsen burner in this procedure, make sure that long hair and loose clothing have been confined.

HRW material copyrighted under notice appearing earlier in this work.

MATERIALS

- 8 oz foam cup
- 100 mL graduated cylinder
- 400 mL beaker
- Bunsen burner and related equipment
- crucible tongs
- ice

- glass stirring rod
- iron ring
- ring stand
- thermometer, nonmercury, 0–100°C
- wire gauze, ceramic-centered

PROCEDURE

1. Heat approximately 120 mL of water to about 45°C in a 400 mL beaker.

2. Using the graduated cylinder, measure 100 ± 1 mL of the warm water and pour the water into the foam cup. Record the volume of water to the nearest 1 mL and the temperature of the water to the nearest 0.1°C in your data table.

3. Using tongs, shake several ice cubes to remove any excess water. Place the ice cubes into the warm water in the foam cup. Use a stirring rod to stir the ice-water mixture until the temperature is less than 1°C. Record the lowest temperature reached to the nearest 0.1°C.

4. Use tongs to remove the unmelted ice. Allow any water on the ice to drip back into the cup. Measure the volume of water in the foam cup to the nearest 1 mL.

Cleanup and Disposal

5. Clean all apparatus and your lab station. Return equipment to its proper place. Wash your hands thoroughly after all work is finished and before you leave the lab.

Data Table		
Volume of warm water	100	mL
Temperature of warm water	50.0	°C
Lowest temperature of ice-water mixture	1.0	°C
Volume of water and melted ice	161	mL

CALCULATIONS

1. **Organizing Data** Calculate the volume of water that came from the melted ice. Record this and all subsequent calculations in your calculations table.

   volume of water from ice = **161 mL − 100 mL = 61 mL**

2. **Organizing Data** Determine the mass of the water that came from the melted ice. Assume the density of water to be 1.00 g/mL.

   mass of ice = **density** × **volume** = **1.00 g/mL** × **61 mL = 61 g**

3. **Organizing Data** Calculate the change in temperature, Δt, of the warm water.

   Δt = **50.0°C − 1.0°C = 49.0°C**

4. **Organizing Data** Determine the mass of the warm water.

mass of warm water = $\underline{1.00 \text{ g/mL} \times 100 \text{ mL} = 100. \text{ g}}$

5. **Organizing Data** Calculate the energy released by the warm water as it cooled through Δt degrees. (Hint: The specific heat of water is 4.184 J/g $\cdot$ °C.)

energy released = $\underline{100. \text{ g} \times 49.0°C \times 4.184 \dfrac{J}{g \cdot °C} = 20\ 500 \text{ J}}$

6. **Organizing Data** Calculate the energy released by the warm water for every gram of ice melted.

energy released/gram of ice melted = $\underline{\dfrac{20\ 500 \text{ J}}{61 \text{ g}} = 340 \text{ J/g}}$

7. **Inferring Conclusions** Calculate the molar heat of fusion of ice (the kilojoules required to melt one mole of ice).

molar heat of fusion = $\underline{\dfrac{340 \text{ J}}{g} \times \dfrac{18.0 \text{ g}}{mol} \times \dfrac{1 \text{ kJ}}{1000 \text{ J}} = 6.12 \text{ kJ/mol}}$

8. **Evaluating Conclusions** The accepted value for the molar heat of fusion of ice is 6.02 kJ. What is the error and percent error in the value you found?

Answers will vary.
error = experimental value − accepted value
$\qquad\qquad\qquad$ = 6.12 kJ/mol − 6.02 kJ/mol = 0.10 kJ/mol
Percent error = $\dfrac{\text{error}}{\text{accepted value}} \times 100$

$\qquad\qquad\qquad\qquad = \dfrac{0.10 \text{ kJ/mol}}{6.02 \text{ kJ/mol}} \times 100 = 1.7\%$

QUESTIONS

1. **Analyzing Ideas** What happens to the kinetic energy of the ice during the process of melting? Explain your answer.

The kinetic energy stays the same because the temperature is constant while melting takes place and kinetic energy varies directly with temperature.

2. **Analyzing Ideas** What happens to the potential energy of ice while it melts? Explain your answer.

<u>Potential energy increases because the water molecules are pulled</u>

<u>apart, in opposition to the attractive forces that hold them</u>

<u>together in the solid.</u>

_____

_____

**GENERAL
CONCLUSIONS**

1. **Applying Conclusions** Before a predicted frost, orange trees are often sprayed with water in an attempt to keep the fruit from freezing. Explain how this could help.

<u>Fruit freezes at a lower temperature than water does. As the</u>

<u>sprayed water freezes, the heat of fusion is released. The released</u>

<u>heat warms the surrounding air to an extent that the air tempera-</u>

<u>ture may not drop below the freezing temperature of the fruit.</u>

_____

_____

Name _____

Date _____ Class _____

EXPERIMENT **A11**

Ice-Nucleating Bacteria

OBJECTIVES

Recommended time:
Two 50-minute periods

- **Observe** the effects of ice-nucleating proteins on ice formation, the freezing temperature of water, and the heat of crystallization.
- **Calculate** the cooling rate for water.
- **Graph** cooling curve data.

INTRODUCTION

Materials

Materials for this activity can be purchased from WARD'S Natural Science Establishment, Inc.
5100 W. Henrietta Road
P.O. Box 92912
Rochester, NY 14692-9012

A common misconception is that water freezes at 0°C under all conditions. In fact, freezing rarely starts at 0°C. Water in its purest state can be "supercooled" to as low as −40°C without ice formation.

An ice nucleator helps water to freeze by attracting the water molecules and slowing them down. A **nucleator** is any foreign particle in water that allows the freezing process to begin. The ice-nucleating protein (INP) that will be used in this investigation is derived from the naturally occurring bacterium *Pseudomonas syringae*. This form of *Pseudomonas* is sometimes called the ice-plus variety because it contains a gene that promotes the formation of proteins that serve as nucleators.

SAFETY

Ice-nucleating protein (order number 36 T 5556)
or Ice-Nucleating Bacteria Study Kit (85 T 3501)

Solution/Material Preparation

1. The 30% CaCl₂ solution is prepared by dissolving 30 g of CaCl₂ in 70 g (mL) of water.

  Always wear safety goggles and a lap apron to protect your eyes and clothing. If you get a chemical in your eyes, immediately flush the chemical out at the eyewash station while calling to your teacher. Know the location of the emergency lab shower and the eyewash station and the procedure for using them.

 Do not touch any chemicals. If you get a chemical on your skin or clothing, wash the chemical off at the sink while calling to your teacher. Make sure you read all labels carefully and follow the precautions on all containers of chemicals you use. If there are no precautions, ask your teacher what precautions to follow. Never return leftover chemicals to their original containers.

 Notify your teacher promptly of any broken glass or cuts. Do not clean up broken glass or spills unless your teacher tells you to do so.

MATERIALS

Beakers for the ice bath should be 600 mL capacity or greater. Using crushed ice makes it easier to push the test tubes into the cooling bath.

- wax pencil or felt-tip marking pen
- test tubes, 3, and rack
- small test tubes, 4
- rubber stopper
- distilled water
- ice-nucleating protein granules
- "ice-nucleating protein-treated water"
- aluminum foil (4 in. × 6 in.)

- graduated plastic pipets, 4
- stopwatch or clock with second hand
- 30% CaCl₂ solution
- ice
- large beaker for ice bath
- cardboard strips, 1/2 in. × 2 in., 2
- − 30°C to 50°C thermometer, 2
- stapler

HRW material copyrighted under notice appearing earlier in this work.

PROCEDURE

2. Prepare separate containers for the disposal of broken glass and liquids.

3. To reduce the amount of time required to complete this investigation, students may start with chilled distilled water of about 10°C (50°F). Treat the water with INP just prior to use. The temperature of both tubes at the beginning of the investigation (t_0) should be the same.

Required Precautions

Discuss all safety precautions with students before the lab.

Disposal

Dilute solutions containing ice-nucleating protein in a ratio of 1 part solution to 20 parts water. Pour the mixture down the drain.

Part 1—Observing the Effects of Ice-Nucleating Protein

1. Use a wax pencil or felt-tip marking pen to label a test tube "INP-treated water."

2. Add 10 mL of distilled water to the test tube. Then add 3 to 4 granules of the ice-nucleating protein to the test tube. Put a rubber stopper in the mouth of the test tube. Mix the contents well by inverting the test tube several times.

3. Make an cooling bath by filling the large beaker three-fourths full with ice. Add 30% $CaCl_2$ until the liquid is at the same level as the ice.

4. Using a graduated plastic pipet, place one drop of the INP-treated water into two small test tubes. *Note: For the best results, keep the water droplets as small as possible.* Save the remaining INP-treated water for later use.

5. With a clean, graduated plastic pipet, place one small drop of distilled water into two small test tubes. *Note: Again, keep the water droplets as small as possible.*

6. Place your test tubes in the cooling bath. Observe the contents of the test tubes every 3 minutes for about 15 minutes or until one set of droplets freezes. Watch for any changes in freezing between the plain distilled water droplets and the INP-treated water droplets. Did the INP-treated water droplets freeze faster than the plain water droplets? Explain your answer.

INP-treated water should freeze faster than plain water because

of the ability of the ice-nucleating protein to start ice crystal

formation at a higher temperature than otherwise possible.

_____

_____

_____

_____

Part 2—Measuring the Effects of INP on Freezing Temperature, Cooling Rate, and Heat of Crystallization

7. Use a wax pencil to label two test tubes *Tube 1* and *Tube 2*. Also label each tube with your initials.

8. Use a clean, graduated plastic pipet to transfer 3 mL of the INP-treated water to *Tube 1*. Place the tube in a test-tube rack.

9. Using a clean, graduated plastic pipet, add 3 mL of distilled water to *Tube 2*, and place the tube in the rack next to *Tube 1*.

10. Fold the cardboard strips in half lengthwise. Place the fold of one piece of cardboard over the top of one of the thermometers. Staple each end to hold the cardboard securely to the thermometer as shown in Figure A. *Note: Be careful not to break the thermometer with the stapler. Repeat for the second thermometer.* When finished stapling, you should be able to slide the cardboard up and down the thermometer.

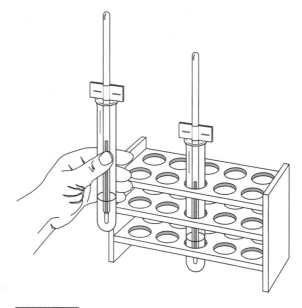

|FIGURE A

11. Insert a thermometer into each test tube. Slide the cardboard strips to a position on each thermometer so that it is suspended in the solution and does not touch the glass. Put the test tubes in the cooling bath.

12. Take temperature readings of *Tube 1* and *Tube 2* at 5-minute intervals for a minimum of 50 minutes. Do not touch the thermometers or tubes. Make careful observations of any ice formation during this period. Record the temperature readings and observations of the test tubes in the data tables.

Table 1	INP-Treated Water		
Time (min)	Temp. Tube 1 (°C)	Cooling rate (°C/min)	Observations
0	18	NA	
5	8	− 2.0	
10	1	− 1.4	Ice starting to form
15	0	− 0.2	
20	0	0	
25	0	0	
30	0	0	
35	0	0	
40	0	0	Solid ice
45	− 1	− 0.2	
50	− 1	0	
55	− 2	− 0.2	
60	− 2	0	

Table 2   Distilled Water

Time (min)	Temp. Tube 2 (°C)	Cooling rate (°C/min)	Observations
0	18	NA	
5	7	− 2.2	
10	2	− 1.0	
15	0	− 0.4	
20	− 1	− 0.2	
25	− 3	− 0.4	Ice starting to form
30	− 2	− 0.2	
35	0	+0.4	
40	− 1	− 0.2	
45	− 1	0	
50	− 2	− 0.2	
55	− 2	− 0.2	Solid ice
60	− 3	− 0.2	

13. Calculate the rate of cooling for each test tube using the following equation. Record your findings in the appropriate data table above.

$$\frac{°C}{min} = \frac{T_2 - T_2}{t_2 - t_1}$$

where

T_1 = temperature at *start* time
T_2 = temperature at *stop* time
t_1 = time at *start*
t_2 = time at *stop*

14. On the grid provided, graph temperature versus time during freezing for *Tube 1* and *Tube 2*. From the graph, determine the heat of crystallization. The heat of crystallization is the amount of heat released when a liquid is transformed into ice crystals. The heat of crystallization for this lab is found by finding the lowest temperature at which water starts to freeze and the highest temperature at which freezing is completed. Find the difference between the two temperatures. Multiply that number by 18 to determine the heat of crystallization.

Cleanup and Disposal

15. Dispose of your materials according to the directions from your teacher. Clean up your work area, and wash your hands before leaving the lab.

ChemFile

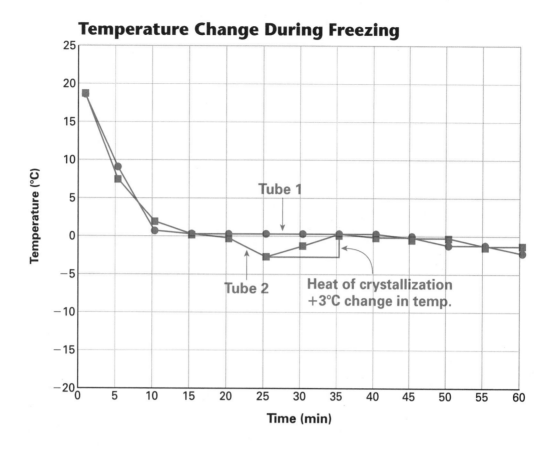

Temperature Change During Freezing

1. **Analyzing Results** Which of the two test tubes from step 12 first started to display signs of ice crystal formation?

   Test-tube 1 containing the INP-treated water should display the

   first signs of ice formation.

2. **Evaluating Data** At what temperature did the distilled water and the INP-treated water begin to freeze? Were these temperatures what you expected them to be?

   The first signs of ice formation of the plain distilled water were

   evident at about $-3°C$, and the INP-treated water started to dis-

   play signs of freezing at about 1°C. No, most would expect water

   to begin freezing at 0°C.

3. Explain the differences in the temperatures for the first signs of freezing.

For freezing to begin, there must be a nucleation site for ice

crystals to grow. INP provides the initial nucleation site for ice

crystal formation. Once ice formation has begun, ice itself can

serve as a nucleator, and thus INP would not be needed after the

initial ice formation stage. Distilled water requires a lower tem-

perature than INP-treated water to start freezing because it lacks

the nucleation sites provided by the INP molecules.

**GENERAL
CONCLUSIONS**

1. Relating Ideas According to your data and graph, what effect did the INP have on the formation of ice?

The temperature at which ice formed in the INP-treated water

was higher (1°C) than that at which ice formed in distilled water

(-3°C).

2. Applying Conclusions How does your result apply to the snow-making operation at the resort?

By adding ice-plus bacteria to the water used in the snow-making

operation, you can now make snow at temperatures higher than

0°C.

Extensions

1. If the temperature outdoors is cold enough, find out if the water solution with the bacteria freezes at the same temperature outside as it did in the lab.

2. Find out how ice-nucleating proteins are used in nature by animals such as insects, amphibians, and reptiles to help them survive freezing during the winter.

Name _____

Date _____ Class _____

EXPERIMENT **A12**

Hydronium Ion Concentration and pH

OBJECTIVES

Recommended time:
60 min

- **Use** pH paper and standard colors to determine the pH of a solution.
- **Determine** hydronium ion concentrations from experimental data.
- **Describe** the effect of dilution on the pH of acids and sodium hydroxide.
- **Relate** pH to the acidity and basicity of solutions.

INTRODUCTION

Solution/Material Preparation

- Wear safety goggles, a face shield, impermeable gloves, and a lab apron when you prepare the NaOH, HCl, H_3PO_4, NH_3, and CH_3COOH solutions. For all except NaOH and H_3PO_4, work in a hood known to be in operating condition, with another person present nearby to call for help in case of an emergency. Be sure you are within a 30 s walk from a safety shower and eyewash station known to be in good operating condition.

1. To prepare 1 L of 0.033 M phosphoric acid, observe the required safety precautions. Slowly and with stirring, add 2 mL of concentrated H_3PO_4 to enough distilled water to make 1.0 L of solution.

2. To prepare 1 L of 0.1 M acetic acid, observe the required safety precautions. Slowly and with stirring, add 6 mL of glacial acetic acid to enough distilled water to make 1.0 L of solution.

For pure water at 25°C, the hydronium ion concentration is 1.0×10^{-7} M or 10^{-7} M.

In acidic solutions, the hydronium ion concentration is *greater* than 1×10^{-7} M. For example, the H_3O^+ concentration of a 0.00 001 M hydrochloric acid (HCl) solution is 1×10^{-5} M.

In basic solutions, the hydronium ion concentration is *less* than 1×10^{-7} M. The hydronium ion concentration in a basic solution can be determined from the equation for K_w.

$$K_w = [H_3O^+][OH^-] = 1 \times 10^{-14} \text{ mol}^2/L^2$$

$$[H_3O^+] = \frac{1 \times 10^{-14} \text{ mol}^2/L^2}{[OH^-]}$$

Therefore, the concentration of H_3O^+ in a 0.000 01 M NaOH basic solution is

$$[H_3O^+] = \frac{1 \times 10^{-14} \text{ mol}^2/L^2}{[OH^-]} = \frac{1 \times 10^{-14} \text{ mol}^2/L^2}{0.000\,01 \text{ mol}/L} = \frac{1 \times 10^{-14} \text{ mol}^2/L^2}{1 \times 10^{-5} \text{ mol}/L}$$
$$= 1 \times 10^{-9} \text{ mol}/L$$

The pH of a solution is defined as the common logarithm of the inverse (reciprocal) of the hydronium ion concentration, $[H_3O^+]$. For a 1×10^{-5} M HCl solution,

$$pH = \log\left(\frac{1}{1 \times 10^{-5}}\right) = \log\,(1 \times 10^5) = 5$$

For a 1×10^{-5} M NaOH solution,

$$[H_3O^+] = \frac{1 \times 10^{-14} \text{ mol}^2/L^2}{1 \times 10^{-5} \text{ mol}/L} = 1 \times 10^{-9} \text{ mol}/L$$

$$\text{and } pH = \log\left(\frac{1}{1 \times 10^{-9}}\right) = \log\,(1 \times 10^{-9}) = 9$$

Values for pH greater than 7 indicate a basic solution. The higher the pH above the value of 7, the stronger the base and the smaller the hydronium ion concentration.

SAFETY

3. To prepare 1 L of 0.1 M hydrochloric acid, observe the required safety precautions. Slowly and with stirring, add 9 mL of concentrated HCl to enough distilled water to make 1.0 L of solution.

4. To prepare 1 L of 0.1 M sodium chloride, dissolve 6 g of NaCl in enough distilled water to make 1.0 L of solution.

5. To prepare 1 L of 0.1 M sodium carbonate, dissolve 5 g of Na_2CO_3 in enough distilled water to make 1.0 L of solution.

Always wear safety goggles and a lab apron to protect your eyes and clothing. If you get a chemical in your eyes, immediately flush the chemical out at the eyewash station while calling to your teacher. Know the location of the emergency lab shower and the eyewash station and the procedure for using them.

Do not touch any chemicals. If you get a chemical on your skin or clothing, wash the chemical off at the sink while calling to your teacher. Make sure you carefully read the labels and follow the precautions on all containers of chemicals that you use. If there are no precautions stated on the label, ask your teacher what precautions you should follow. Do not taste any chemicals or items used in the laboratory. Never return leftovers to their original containers; take only small amounts to avoid wasting supplies.

Call your teacher in the event of a spill. Spills should be cleaned up promptly, according to your teacher's directions.

Never put broken glass into a regular waste container. Broken glass should be disposed of properly.

MATERIALS

6. To prepare 1 L of 0.1 M sodium hydrogen carbonate, dissolve 8 g of $NaHCO_3$ in enough distilled water to make 1.0 L of solution.

- 0.033 M H_3PO_4
- 0.10 M CH_3COOH
- 0.10 M HCl
- 0.10 M NaCl
- 0.10 M Na_2CO_3
- 0.10 M $NaHCO_3$
- 0.10 M NaOH
- 0.10 M NH_3, aqueous
- 0.10 M NH_4CH_3OO

- 10 mL graduated cylinder
- 50 mL graduated cylinder
- 250 mL beaker
- distilled water
- glass plate
- glass stirring rod
- pH papers, wide and narrow range
- test tubes, 12
- white paper, small sheet

PROCEDURE

7. To prepare 1 L of 0.1 M sodium hydroxide, observe the required safety precautions. Slowly and with stirring, dissolve 4 g of NaOH in enough distilled water to make 1.0 L of solution.

8. To prepare 1 L of 0.1 M aqueous ammonia, observe the required safety precautions. Slowly and with stirring, add 7 mL of concentrated $NH_3(aq)$ to enough distilled water to make 1.0 L of solution.

1. Place the glass plate on the sheet of white paper. Place a strip of wide-range pH paper and a strip of narrow-range pH paper on the glass plate.

2. Obtain samples of all of the solutions listed in Data Table 1. To find the pH of each solution, dip a clean stirring rod into each solution and apply a drop of the solution first to the wide-range pH paper and then to the narrow-range paper. Figure A shows the correct technique. Compare the color produced by each solution with the colors on the chart included with the pH paper. Be sure to rinse and dry the stirring rod before you test each solution. Record your results in Data Table 1.

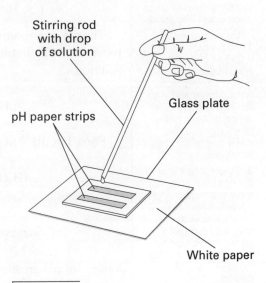

Stirring rod with drop of solution

pH paper strips

Glass plate

White paper

FIGURE A

9. To prepare 1 L of 0.1 M ammonium acetate, dissolve 8 g of NH_4CH_3OO in enough distilled water to make 1.0 L of solution.

10. Best results are obtained by using Panapeha brand strips. Each strip contains six squares for color matching and covers pH values of 1–14. Also available are EM brand strips, which contain four squares per strip. EM paper is available in wide and narrow range.

Required Precautions

• Read all safety pre-cautions, and discuss them with your students.

• Safety goggles and a lab apron must be worn at all times.

• In case of an acid or base spill, first dilute with water. Then, mop up the spill with wet cloths or a wet cloth mop while wearing disposable plastic gloves. Designate sep-arate cloths or mops for acid and base spills.

Procedural Tips

• Show students how to test the solutions and compare the test paper with the pH wide-range and narrow-range stan-dards. Caution them to avoid poor results by thoroughly clean-ing and drying the stirring rod between tests.

3. Using the graduated cylinders, measure 5.0 mL of 0.10 M HCl and dilute it to 50 mL with water. Transfer 5 mL of the diluted solution to a labeled test tube.

4. Again using the graduated cylinders, dilute 5.0 mL of the diluted solution in step **3** to 50 mL, and transfer 5 mL of this solution to another labeled test tube. Repeat this procedure two more times. When completed, you should have four solutions of HCl with concentrations of 0.10 M, 0.010 M, 0.0010 M, and 0.000 10 M.

5. Repeat steps **3** and **4** for 0.10 M NaOH and 0.10 M CH_3COOH. Record your results for each test in Data Table 2.

Cleanup and Disposal

6. Clean all apparatus and your lab station. Return equipment to its proper place. Dispose of chemicals and solutions in the containers designated by your teacher. Do not pour any chemicals down the drain or in the trash unless your teacher directs you to do so. Wash your hands thoroughly after all work is finished and before you leave the lab.

Data Table 1

0.1 M solution	pH	0.1 M solution	pH
HCl	1	NaCl	7
CH_3COOH	3	Na_2CO_3	11
H_3PO_4	1.5	$NaHCO_3$	8
NaOH	13	NH_4CH_3COO	7
NH_3	11		

Data Table 2

Concentration (M)	HCl	NaOH	CH_3COOH
0.10	Students' answers will vary.		
0.010			
0.0010			
0.000 10			

QUESTIONS

1. Organizing Data List the solutions in order of decreasing acid strength using your results from step **2.**

HCl, H_3PO_4, CH_3COOH, NH_4CH_3COO, NaCl, $NaHCO_3$, NH_3,

Na_2CO_3, NaOH

• Combine all solutions containing H_3PO_4 and adjust the pH to between 5 and 9 using 1.0 M KOH. (Do not use NaOH because the EPA does not want sodium hydrogen phosphates in the water.) Then pour down the drain. Combine all other solutions and adjust the pH to be between 6 and 8 and pour down the drain.

2. Organizing Data Calculate the theoretical pH values for the concentrations prepared in steps **3–5**. Record these values below.

Calculated pH

HCl		NaOH	
0.1 M	1	0.1 M	13
0.010 M	2	0.010 M	12
0.0010 M	3	0.0010 M	11
0.000 10 M	4	0.000 10 M	10

3. Analyzing Results What effect does dilution have on the pH of (a) an acid and (b) a base?

a. The pH of an acid increases as the solution is made more dilute.

b. The pH of a base decreases as the solution is made more dilute.

GENERAL CONCLUSIONS

1. Predicting Outcomes Solutions with a pH of 12 or greater dissolve hair. Would a cotton shirt or a wool shirt be affected more by a spill of 0.1 M sodium hydroxide? Explain.

A shirt made from wool, which is animal hair, will dissolve in the

sodium hydroxide solution.

Name _____

Date _____ Class _____

EXPERIMENT **A13**

Titration of an Acid with a Base

OBJECTIVES

Recommended time:
60 min

- **Use** burets to accurately measure volumes of solution.
- **Recognize** the end point of a titration.
- **Describe** the procedure for standardizing a solution.
- **Determine** the molarity of a base.

INTRODUCTION

Solution/Material Preparation

1. To prepare 1 L of 0.500 M hydrochloric acid, use a fresh bottle of reagent grade concentrated HCl, preferably one that shows the actual assay of HCl rather than an average assay. Observe the required safety precautions. Assuming that the concentrated HCl is 12 M, slowly and with stirring, add 41.65 mL to enough distilled water to make 1.00 L of solution. If a large number of students are to be provided with solutions, it's best to make up 10 L in a large dispenser to ensure the same concentration for all lab groups. However, it will be impossible to make the HCl exactly 0.500 M without the accuracy of a volumetric flask. Five-gallon carboys with spigots make ideal dispenser for the acid and base solutions.

Titration is a process in which the concentration of a solution is determined by measuring the volume of that solution needed to react completely with a standard solution of known volume and concentration. The process consists of the gradual addition of the standard solution to a measured quantity of the solution of unknown concentration until the number of moles of hydronium ion, H_3O^+, equals the number of moles of hydroxide ion, OH^-. The point at which equal numbers of moles of acid and base are present is known as the equivalence point. An indicator is used to signal when the equivalence point is reached. The chosen indicator must change color very near or at the equivalence point. The point at which an indicator changes color is called the end point of the titration. Phenolphthalein is an appropriate choice for this titration. In acidic solution, phenolphthalein is colorless, and in basic solution, it is pink.

At the equivalence point, the number of moles of acid equals the number of moles of base.

$$(1) \quad \text{moles of } H_3O^+ = \text{moles of } OH^-$$

By definition

$$(2) \quad \text{molarity (moles/L)} = \frac{\text{moles}}{\text{volume (L)}}$$

If you rearrange equation 2 in terms of moles, equation 3 is obtained.

$$(3) \quad \text{moles} = \text{molarity (moles/L)} \times \text{volume (L)}$$

When equations 1 and 3 are combined, you obtain the relationship that is the basis for this experiment, assuming a one-to-one mole ratio.

$$(4) \quad \text{molarity of acid} \times \text{volume of acid} = \text{molarity of base} \times \text{volume of base}$$

In this experiment, you will be given a standard HCl solution and told what its concentration is. You will carefully measure a volume of it and determine how much of the NaOH solution of unknown molarity is needed to neutralize the acid sample. Using the data you obtain and equation 4, you can calculate the molarity of the NaOH solution.

SAFETY

2. To prepare 1 L of 0.6 M sodium hydroxide solution, observe the required safety precautions. Add 24 g of NaOH with stirring to enough distilled water to make 1.0 L of solution.

3. To prepare 1 L of phenolphthalein solution, dissolve 10 g of phenolphthalein in 500 mL of denatured alcohol and add 500 mL of water.

Required Precautions

• Read all safety precautions, and discuss them with your students.

• Safety goggles and a lab apron must be worn at all times.

MATERIALS

• In case of an acid or base spill, first dilute with water. Then, mop up the spill with wet cloths or a wet cloth mop while wearing disposable plastic gloves. Designate separate cloths or mops for acid and base spills.

PROCEDURE

• Wear safety goggles, a face shield, impermeable gloves, and a lab apron when you prepare the NaOH and HCl solutions. For preparing HCl, work in a hood known to be in operating condition and have another person stand by to call for help in case of an emergency. Be sure you are within a 30 s walk from a safety shower and eyewash station known to be in good operating condition.

  Always wear safety goggles and a lab apron to protect your eyes and clothing. If you get a chemical in your eyes, immediately flush the chemical out at the eyewash station while calling to your teacher. Know the location of the emergency lab shower and the eyewash station and the procedure for using them.

 Do not touch any chemicals. If you get a chemical on your skin or clothing, wash the chemical off at the sink while calling to your teacher. Make sure you carefully read the labels and follow the precautions on all containers of chemicals that you use. If there are no precautions stated on the label, ask your teacher what precautions you should follow. Do not taste any chemicals or items used in the laboratory. Never return leftovers to their original containers; take only small amounts to avoid wasting supplies.

 Call your teacher in the event of a spill. Spills should be cleaned up promptly, according to your teacher's directions.

 Never put broken glass into a regular waste container. Broken glass should be disposed of properly.

• 0.500 M HCl
• 50 mL burets, 2
• 100 mL beakers, 3
• 125 mL Erlenmeyer flask
• double buret clamp
• NaOH solution of unknown molarity
• phenolphthalein indicator
• ring stand
• wash bottle

1. Set up the apparatus as shown in Figure A. Label the burets *NaOH* and *HCl*. Label two beakers *NaOH* and *HCl*. Place approximately 80 mL of the appropriate solution into each beaker.

2. Pour 5 mL of NaOH solution from the beaker into the NaOH buret. Rinse the walls of the buret thoroughly with this solution. Allow the solution to drain through the stopcock into another beaker and discard it. Rinse the buret two more times in this manner, using a new 5 mL portion of NaOH solution each time. Discard all rinse solutions.

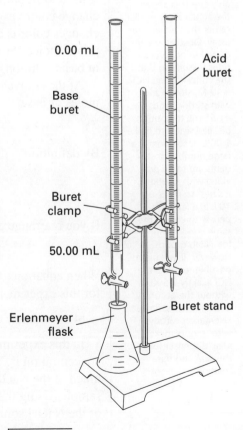

0.00 mL

Base buret

Acid buret

Buret clamp

50.00 mL

Erlenmeyer flask

Buret stand

FIGURE A

• In case of an acid or base spill, dilute first with water. Then, mop up the spill with wet cloths designated for spill cleanup while wearing disposalable plastic gloves. A wet cloth mop can be rinsed out a few times and used until it falls apart.

Procedural Tips

• Demonstrate all techniques needed for successful titration: cleaning the burets, reading the buret with the eye at the liquid level, reading the buret scale correctly, swirling the flask, manipulating the stopcock, washing down the sides of the flask, and evaluating the color of the indicator.

• Discuss the role of the indicator, and the meaning of the terms *end point* and *equivalence point.* Indicators change colors at different pH values. It is important to chose an indicator that changes color at a pH which is close to the pH of the equivalence point of the titration. For strong bases, such as NaOH, titrated with strong acids, such as HCl, phenolphthalein is a good indicator because the pH of its end point is very close (near pH 7) to the equivalence point.

• It may help students if you work through a set of sample titration data.

3. Fill the buret with NaOH solution above the zero mark. Withdraw enough solution to remove any air from the buret tip, and bring the liquid level down within the graduated region of the buret.

4. Repeat steps **2** and **3** with the HCl buret, using HCl solution to rinse and fill it.

5. For trial 1 record the initial reading of each buret, estimating to the nearest 0.01 mL in the Data Table. For consistent results, have your eyes level with the top of the liquid each time you read the buret. Always read the scale at the bottom of the meniscus.

6. Draw off about 10 mL of NaOH solution into an Erlenmeyer flask. Add some distilled water to increase the volume. Add one or two drops of phenolphthalein solution as an indicator.

7. Begin the titration by slowly adding HCl from the buret to the Erlenmeyer flask while mixing the solution by swirling it as shown in Figure B. Stop frequently and wash down the inside surface of the flask using your wash bottle.

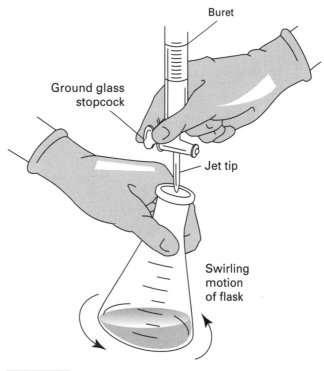

FIGURE B

8. When the pink color of the solution begins to disappear at the point of contact with the acid, add the acid drop by drop, swirling the flask gently after each addition. When the last drop added causes the color to disappear from the whole solution and the color does not reappear, stop the titration. A white sheet of paper under the Erlenmeyer flask makes it easier to detect the color change.

Disposal

Set out three disposal containers for the students: one for unused acid solutions, one for unused base solutions, and one for partially neutralized substances and the contents of the waste beaker. One at a time, slowly combine the solutions while stirring. Adjust the pH of the final waste liquid with 1.0 M acid or base until the pH is between 5 and 9. Pour the neutralized liquid down the drain.

9. Add NaOH solution dropwise just until the pink color returns. Add HCl again, dropwise, until the color just disappears. Go back and forth over the end point several times until one drop of the basic solution just brings out a faint pink color. Wash down the inside surface of the flask and make dropwise additions, if necessary, to reestablish the faint pink color. Read the burets to the nearest 0.01 mL, and record these final readings in your data table.

10. Discard the liquid in the flask, rinse thoroughly with distilled water, and run a second and third trial.

11. Record the known concentration of the standard HCl solution in the data table.

Cleanup and Disposal

12. Clean all apparatus and your lab station. Return equipment to its proper place. Dispose of chemicals and solutions in the containers designated by your teacher. Do not pour any chemicals down the drain or in the trash unless your teacher directs you to do so. Wash your hands thoroughly before you leave the lab and after all work is finished.

Data Table

Buret readings (mL)

Trial	HCl Initial	HCl Final	NaOH Initial	NaOH Final
1	0.70	10.90	5.08	13.58
2	10.90	20.80	13.58	21.80
3	20.80	28.81	21.80	28.33

Molarity of HCl 0.500 M

CALCULATIONS

1. **Organizing Data** Calculate the volumes of acid used in the three trials. Show your calculations.

Trial 1: Volume of HCl = 10.90 mL − 0.70 mL = 10.20 mL

Trial 2: Volume of HCl = 20.80 mL − 10.90 mL = 9.90 mL

Trial 3: Volume of HCl = 28.81 mL − 20.80 mL = 8.01 mL

2. **Organizing Data** Calculate the volumes of base used in the three trials. Show your calculations.

Trial 1: Volume of NaOH = 13.58 mL − 5.08 mL = 8.50 mL

Trial 2: Volume of NaOH = 21.80 mL − 13.58 mL = 8.22 mL

Trial 3: Volume of NaOH = 28.33 mL − 21.80 mL = 6.53 mL

3. **Organizing Data** Use equation 3 in the Introduction to determine the moles of acid used in each of the three trials.

Trial 1: Moles of acid = **0.500 M × 0.01020 L = 0.00510 mol**

Trial 2: Moles of acid = **0.500 M × 0.00990 L = 0.00495 mol**

Trial 3: Moles of acid = **0.500 M × 0.00801 L = 0.00401 mol**

4. **Relating Ideas** Write the balanced equation for the reaction between HCl and NaOH.

HCl + NaOH → NaCl + H₂O

5. **Organizing Ideas** Use the mole ratio in the balanced equation and the moles of acid from Calculations item **3** to determine the moles of base neutralized in each trial.

Trial 1: Moles of acid = **0.00510 mol = moles of base**

Trial 2: Moles of acid = **0.00495 mol = moles of base**

Trial 3: Moles of acid = **0.00401 mol = moles of base**

6. **Organizing Data** Use equation 2 in the Introduction and the results of Calculations item **2** and **5** to calculate the molarity of the base for each trial. Record your results.

Trial 1: Molarity of NaOH $= \dfrac{\text{moles NaOH}}{\text{volume NaOH}} = \dfrac{0.00510 \text{ mol}}{0.00850 \text{ L}} = 0.600 \text{ M}$

Trial 2: Molarity of NaOH $= \dfrac{0.00495 \text{ mol}}{0.00822 \text{ L}} = 0.602 \text{ M}$

Trial 3: Molarity of NaOH $= \dfrac{0.00401 \text{ mol}}{0.00653 \text{ L}} = 0.614 \text{ M}$

7. **Organizing Conclusions** Calculate the average molarity of the base. Record your result.

Average molarity of NaOH $= \dfrac{0.600 \text{ M} + 0.602 \text{ M} + 0.614 \text{ M}}{3} = 0.605 \text{ M}$

**GENERAL
CONCLUSIONS**

1. **Analyzing Methods** In step **6,** you added distilled water to the NaOH
   solution in the Erlenmeyer flask before titrating. Why did the addition of the
   water not affect the results?

   Diluting with water did not change the number of moles of NaOH

   in the flask. HCl was added until the number of moles of HCl

   equaled the number of moles of NaOH.

   _____

   _____

2. **Analyzing Methods** What characteristic of phenolphthalein made it appropriate to use in this titration? Could you have done the experiment without
   it? How does phenolphthalein's end point relate to the equivalence point of
   the reaction?

   Phenolphthalein changes from colorless in acid solution to pink in

   basic solution. Without an indicator such as phenolphthalein,

   there would have been no visual way to determine when the

   equivalence point had been reached. Phenolphthalein's end point

   is close to the equivalence point (neutral pH) of this titration.

   _____

   _____

   _____

EXPERIMENT **A14**

Energy and Entropy

OBJECTIVES

Recommended time:
60 min

- **Observe** the temperature and phase changes of a pure substance.
- **Describe** the relationship between heat energy and phase changes.
- **Graph** experimental data and **analyze** the relationship between time and temperature.

INTRODUCTION

Solution/Material Preparation

1. Have 15 g of sodium thiosulfate pentahydrate available to students in the 25 ×100 mm test tubes. Weigh out one sample, and then fill the remaining test tubes up to the level of the first one.

2. Have seeds crystals of sodium thiosulfate pentahydrate available in a wide-mouthed bottle.

As heat energy flows from a liquid, the temperature of the liquid drops. The entropy (degree of randomness, or disorder, of particles in the liquid) also decreases. If energy continuously flows from a liquid, the liquid eventually will undergo a phase change to a solid. While the phase change is occurring, the temperature of a pure substance will not change. Entropy, however, will continue to decrease. Once the phase change is complete, the temperature of the solid will begin to decrease again if energy continues to be removed.

Liquid water placed in the freezer compartment of a refrigerator undergoes these changes. Heat flows from the liquid water. The temperature of the water decreases to 0°C and then begins to freeze. While the water is freezing, its temperature remains at 0°C. When the phase change is complete and no liquid water remains, the frozen water, or ice, again begins to decrease in temperature, and it continues to cool until it has the same temperature as the freezer, about −12°C. A refrigerator icemaker utilizes this principle. Ice cubes are ejected only after the water has completely frozen and the temperature reaches −9°C.

In this experiment, you will measure the temperature of sodium thiosulfate pentahydrate as it is cooled to several degrees below its freezing temperature and then warmed to several degrees above its melting temperature. You will determine the freezing and melting temperatures of $Na_2S_2O_3 \cdot 5H_2O$ and interpret the changes in energy and entropy.

SAFETY

Always wear safety goggles and a lab apron to protect your eyes and clothing. If you get a chemical in your eyes, immediately flush the chemical out at the eyewash station while calling to your teacher. Know the location of the emergency lab shower and the eyewash station and the procedure for using them.

Do not touch any chemicals. If you get a chemical on your skin or clothing, wash the chemical off at the sink while calling to your teacher. Make sure you carefully read the labels and follow the precautions on all containers of chemicals that you use. If there are no precautions stated on the label, ask your teacher what precautions you should follow.

3. To make a wire stirrer, use any easily bendable gauge wire. At one end, make a circular loop with a diameter less than the diameter of the test tube. Bend the loop perpendicular to the length of the wire. The wire should be long enough to extend two or three inches above the top of the test tube.

Call your teacher in the event of a spill. Spills should be cleaned up promptly, according to your teacher's directions.

When using a Bunsen burner, confine long hair and loose clothing. If your clothing catches on fire, WALK to the emergency lab shower and use it to put out the fire. Do not heat glassware that is broken, chipped, or cracked. Use tongs or a hot mitt to handle heated glassware and other equipment because hot glassware does not always look hot.

Never put broken glass in a regular waste container. Broken glass should be disposed of properly.

MATERIALS

Required Precautions

• Safety goggles and a lab apron must be worn at all times.

• Tie back long hair and loose clothing when working in the lab.

• Read all safety cautions, and discuss them with your students.

• Make sure the iron rings are large enough to hold a 600 mL beaker.

Procedural Tips

• Discuss the reason for the use of a seed crystal, and ask students to think about this step in the procedure as they process their experimental data.

PROCEDURE

• In Part II, students may have difficulty keeping the temperature of the water bath constant around 60°C. Demonstrate the necessary small size of the burner flame and how to position the burner well off the center of the beaker.

• 600 mL beakers, 2
• balance, centigram
• Bunsen burner and related equipment, or hot plate
• buret clamp
• clock with second hand
• forceps
• iron ring
• $Na_2S_2O_3 \cdot 5H_2O$, sodium thiosulfate pentahydrate
• ring stand
• ruler
• sparker
• test tube, 25 ×100 mm
• test-tube rack
• thermometers, non-mercury, 2
• wire gauze, ceramic-centered
• wire stirrer

Freezing a Liquid

1. Clamp the ring on the ring stand so that it is 10 cm above the top of the burner, as shown in Figure A. Fill a 600 mL beaker three-fourths full with water, and place it on the iron ring and gauze. Heat the water to 85°C. Fill another 600 mL beaker three-fourths full with water to use as the cold-water bath.

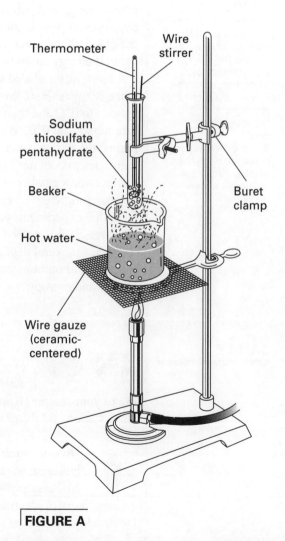

Thermometer — Wire stirrer — Sodium thiosulfate pentahydrate — Beaker — Buret clamp — Hot water — Wire gauze (ceramic-centered)

FIGURE A

• To avoid breakage, caution students not to try to move the thermometer when the sodium thiosulfate pentahydrate is solid. Demonstrate the up and down action of the wire stirrer.

• Encourage students to write down their observations in the data table in order to relate them to the graph of their data and to changes in kinetic energy and entropy.

Disposal

Remelt the sodium thiosulfate pentahydrate in a water bath, pour all of the liquid in a wide-mouth reagent jar, cool to room temperature, cover, and label for reuse. It will be necessary to pulverize the crystals into smaller chunks before reusing them.

2. Obtain a 25 ×100 mm test tube containing approximately 15 g of sodium thiosulfate pentahydrate. Clamp the test tube above the hot-water bath as shown in Figure A.

3. When the temperature of the hot-water bath reaches 85°C, turn off the burner. Immerse the test tube in the hot-water bath. Occasionally stir the melting solid with a wire stirrer. Observe the temperature of the melting solid using one of the thermometers.

 Never stir with a thermometer because the glass around the bulb is fragile, and might break.

4. When the temperature of the liquid sodium thiosulfate is approximately the same as the temperature of the hot-water bath, remove the test tube from the hot water. Record the time and the temperature of the liquid sodium thiosulfate pentahydrate in the data table. Immediately immerse the test tube in the cold-water bath. Stir with the wire stirrer, and record the temperature of the sodium thiosulfate pentahydrate at 15 s intervals.

5. When the temperature reaches 50°C, use forceps to add one or two seed crystals of sodium thiosulfate pentahydrate to the test tube. Continue recording temperature readings every 15 s. Constantly stir the liquid-solid mixture in the test tube until a constant temperature (between 45°C and 50°C) is maintained. Do not try to move the thermometer when solidification occurs.

6. Continue to read and record the temperature of the solid until the temperature is within five degrees of that of the cold-water bath.

Melting a Solid

7. Clamp the test tube above the hot-water bath as shown in Figure A. Heat the water to 60–65°C. Do not allow the temperature of the water to exceed 65°C. Turn the flame down as low as possible, and move the burner so only the outer edge of the bottom of the beaker is heated. Adjust the position and size of the flames so the temperature of the hot-water bath remains between 60 and 65°C.

8. Record the temperature of the sodium thiosulfate pentahydrate and the time in your data table. The starting temperature should be lower than 35°C. Immediately immerse the test tube in the hot-water bath. Record the temperature of the sodium thiosulfate pentahydrate at 15 s intervals. Begin stirring with the wire stirrer as the mixture begins to soften. Continue reading and recording the temperature until it is within five degrees of the temperature of the hot-water bath.

Cleanup and Disposal

9. Clean all lab equipment and your lab station. Return equipment to its proper place. Dispose of chemicals and solutions in the containers designated by your teacher. Do not pour any chemicals down the drain or in the trash unless your teacher directs you to do so. Wash your hands thoroughly after all work is finished and before you leave the lab.

Data Table

Cooling data

Time (s)	Temp (°C)	Observations	Time (s)	Temp (°C)	Observations
0	69.0		210	48.2	
15	61.0		225	48.2	
30	51.0	seed crystal added	240	48.0	total solidification
45	47.0	crystallization begins	255	47.5	
60	46.0		270	47.0	
75	48.2		285	45.7	
90	48.2		300	44.7	
105	48.2		315	43.5	
120	48.2		330	41.5	
135	48.2		345	39.0	
150	48.2		360	36.2	
165	48.2		375	34.0	
180	48.2		390	32.0	
195	48.2				

Data Table

Warming data

Time (s)	Temp (°C)	Observations	Time (s)	Temp (°C)	Observations
0	29.0		210	47.4	
15	31.5		225	47.5	
30	36.0		240	47.5	
45	39.5		255	47.5	
60	42.0		270	48.0	
75	43.5		285	55.0	all solid melted
90	44.5		300	55.0	
105	45.5		315	58.0	
120	46.0	solid melting	330	60.0	
135	46.0		345		
150	46.5		360		
165	47.0		375		
180	47.0		390		
195	47.2				

CALCULATIONS

1. **Analyzing Results** Presenting data graphically makes it easier to compare two sets of data. Plot your warming and cooling data on the grid using the horizontal axis for the time and the vertical axis for the temperature.

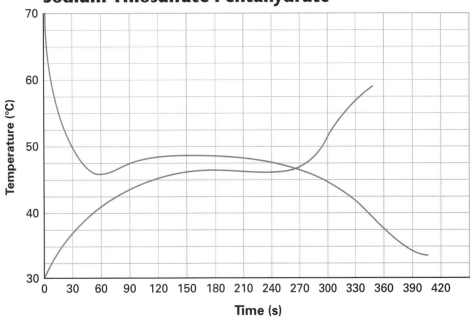

Time-Temperature Graph of Heating and Cooling Sodium Thiosulfate Pentahydrate

QUESTIONS

1. **Evaluating Data** What is the shape of the part of the graph that represents the cooling of the liquid?

   It slopes downward.

2. **Analyzing Ideas** While the liquid was freezing to the solid, what was happening to **a.** the temperature and **b.** the entropy?

   a. The temperature remained relatively constant.

   b. The entropy was decreasing.

3. **Analyzing Results** According to your experimental data, what is the freezing point of sodium thiosulfate pentahydrate?

   48°C (approximately 48.5°C from data)

4. **Analyzing Results**   According to your experimental data, what is the melting point of sodium thiosulfate pentahydrate?

48°C (approximately 49.5°C from data)

5. **Relating Ideas**   What is the relationship between the freezing point and the melting point of a substance?

They are the same temperature.

GENERAL CONCLUSIONS

1. **Relating Ideas**   Refer to the cooling curve you plotted. For each portion of the curve, describe what is happening to the kinetic energy and the entropy of sodium thiosulfate pentahydrate?

In the first 60 s, the kinetic energy and the entropy decrease.

Between 60 s and approximately 270 s, the kinetic energy

remains relatively constant but the entropy decreases. After 270 s,

both the kinetic energy and the entropy decrease.

Date _____ Class _____

EXPERIMENT **A15**

Heat of Crystallization

OBJECTIVES

Recommended time:
40 min

- **Observe** temperature changes in the surroundings of a liquid as it crystallizes.
- **Compute** the amount of energy released from the observed temperature change.
- **Determine** the energy released per gram of sodium thiosulfate pentahydrate.
- **Relate** changes in temperature to changes in kinetic and potential energy.

INTRODUCTION

When a pure substance undergoes a phase change, its temperature remains constant. This does not mean that there are no energy changes taking place. In fact, when a substance such as liquid sodium thiosulfate pentahydrate freezes, the molecules in the liquid phase lose potential energy as they become part of a rigid crystal structure. This loss of potential energy can be observed as an increase in the kinetic energy of the surroundings, that is, an increase in temperature.

In this experiment, you will determine the amount of heat released as supercooled liquid sodium thiosulfate pentahydrate crystallizes. A mass of water will absorb the heat, and from the mass of the water and the temperature change, you will be able to calculate the heat of crystallization of sodium thiosulfate in joules per gram.

SAFETY

Required Precautions

- Safety goggles and a lab apron must be worn at all times.
- Read all safety precautions, and discuss them with your students.
- In case of a spill, use a dampened cloth or paper towels to mop up the spill. Then rinse the cloth in running water at the sink, wring it out until it is damp, and put it in the trash.

Always wear safety goggles, gloves, and a lab apron to protect your eyes, hands, and clothing. If you get a chemical in your eyes, immediately flush the chemical out at the eyewash station while calling to your teacher. Know the location of the emergency lab shower and eyewash station and the procedure for using them.

Do not touch any chemicals. If you get a chemical on your skin or clothing, wash the chemical off at the sink while calling to your teacher. Make sure you carefully read the labels and follow the precautions on all containers of chemicals that you use. If there are no precautions stated on the label, ask your teacher what precautions you should follow. Do not taste any chemicals or items used in the laboratory. Never return leftovers to their original containers; take only small amounts to avoid wasting supplies.

When using a Bunsen burner, confine long hair and loose clothing. If your clothing catches on fire, WALK to the emergency lab shower, and use it to put out the fire. Do not heat glassware that is broken, chipped, or cracked. Use tongs or a hot mitt to handle heated glassware and other equipment because hot glassware and other equipment does not always look hot.

Call your teacher in the event of a spill. Spills should be cleaned up promptly, according to your teacher's directions.

Never put broken glass in a regular waste container. Broken glass should be disposed of properly in the broken glass waste container.

MATERIALS

Procedural Tips

• A review of the principles of calorimetry might be helpful. Ask students why they will be measuring the change in the temperature of the water rather than the liquid sodium thiosulfate.

• Caution students not to allow the temperature of the liquid sodium thiosulfate to rise above 75°C. Also emphasize that the liquid must not be stirred while cooling.

PROCEDURE

• Insist that they use a stirring rod—and not a thermometer—to stir the water bath.

Disposal

• Put out containers lined with plastic wrap for students to dispose of their sodium thiosulfate. The recovered material can be reused.

- 100 mL graduated cylinder
- 400 mL beakers, 2
- balance, centigram
- Bunsen burner and related equipment
- foam cup, 8 oz
- forceps
- glass stirring rod
- iron ring
- $Na_2S_2O_3 \cdot 5H_2O$
- test tube, 25 × 100 mm
- ring stand
- sparker
- test-tube clamp
- thermometers, nonmercury, 2
- wire gauze, ceramic-centered

1. Prepare a boiling-water bath using one of the 400 mL beakers. Fill the other 400 mL beaker with water, and allow it to adjust to room temperature. **CAUTION Before you light the Bunsen burner, check to see that long hair and loose clothing have been confined.**

2. Obtain the mass of approximately 15 g of sodium thiosulfate pentahydrate to the nearest 0.1 g, and record the mass in your data table.

3. Transfer the sodium thiosulfate to the test tube. Clamp the test tube to the ring stand and lower it into the boiling-water bath. Put a thermometer in the test tube.

4. Allow the solid to melt. Continue to heat the water bath until the temperature of the liquid sodium thiosulfate is approximately 75°C.

5. Remove the test tube from the boiling-water bath and place it in the room-temperature water bath you prepared in step **1. Do not stir or shake the liquid in the test tube.**

6. Fill a foam cup about two-thirds full with tap water. Add an ice cube and cool the water until the temperature is 10–15°C. Use a graduated cylinder to measure approximately 75 mL of the cooled water and record the volume to the nearest 0.1 mL in your data table. Discard the rest of the cooled water. Dry the cup and pour the cooled water from the graduated cylinder into it.

7. When the temperature of the liquid sodium thiosulfate has dropped below 35°C, remove the test tube from the room-temperature bath and immediately put it into the cold-water bath. Using forceps, drop a seed crystal into the test tube. Continually stir the cold-water bath with a stirring rod, and carefully

ChemFile

observe how the bath temperature changes. Record these observations in the Data Table.

8. Record to the nearest 0.1°C the highest temperature reached by the cold-water bath (not the sodium thiosulfate).

Cleanup and Disposal

9. Clean all apparatus and your lab station. Return equipment to its proper place. Dispose of chemicals and solutions in the containers designated by your teacher. Do not pour any chemicals down the drain or in the trash unless your teacher directs you to do so. Wash your hands thoroughly after all work is finished and before you leave the lab.

Data Table		
Mass of sodium thiosulfate	15.0	g
Volume of cold-water bath	75.0	mL
Initial temperature of cold-water bath	13.5	°C
Final temperature of cold-water bath	23.5	°C
Observations (step 7)		

CALCULATIONS

1. **Organizing Data** Determine the change in temperature, Δt, of the cold-water bath.

$$\Delta t = 23.5°C - 13.5°C = 10.0°C$$

2. **Organizing Data** Calculate the heat in joules absorbed by the cold-water bath. (Hint: the specific heat of water is 4.184 J/g·°C.)

$$\text{heat absorbed} = \text{mass(g)} \times \Delta t(°C) \times 4.184 \text{ J/g·°C}$$

$$= 75.0 \text{ g} \times 10.0°C \times 4.184 \text{ J/g·°C}$$

$$= 3140 \text{ J}$$

3. **Inferring Conclusions** Calculate the joules of heat released per gram of sodium thiosulfate.

$$\text{heat released/gram} = \frac{3140 \text{ J}}{15.0 \text{ g}} = 209 \text{ J/g}$$

4. **Evaluating Conclusions** The accepted value for the heat of crystallization of sodium thiosulfate is 200. J/g. Calculate your error and percent error.

error = experimental value − accepted value

= 209 J/g − 200 J/g = 9 J/g

percent error = $\dfrac{9 \text{ J/g}}{200 \text{ J/g}} \times 100 = 4.5\%$

**GENERAL
CONCLUSIONS**

1. **Analyzing Information** In step **7,** what is the source of the heat energy
that is released by the sodium thiosulfate pentahydrate and absorbed by the
water?

The molecules in the liquid phase lose potential energy as they

become more rigidly positioned in the solid phase. This energy is

transferred to the water, raising its temperature and the kinetic

energy of the water molecules.

2. **Applying Conclusions** One problem with solar heating is that it is diffi-
cult to store the sun's heat for nighttime heating. How might sodium thiosulfate pentahydrate be used to solve this problem?

The sun's heat could be used during the daytime to melt sodium

thiosulfate pentahydrate crystals. In the evening, when heat is

needed, a seed crystal could be added to the supercooled liquid.

As the liquid crystallizes, heat would be released.

3. **Applying Conclusions** Honey is a supercooled liquid. How could you
crystallize a jar of honey?

Add a seed crystal of sugar from a jar of honey that has already

crystallized to a jar of liquid honey. Slowly, over the next few

months, the clear jar of honey will become cloudy and crystal-

lized.

EXPERIMENT **A16**

Temperature of a Bunsen Burner Flame

OBJECTIVES

Recommended time:
25 min

- **Observe** the change in temperature of a known mass of water when a heated metal object of known mass is placed in it.

- **Use** the specific heat of the metal and of water to calculate the initial temperature of the object.

- **Relate** the temperature of the object to the temperature of the Bunsen burner flame.

INTRODUCTION

Solution/Material Preparation

1. Large copper or iron nuts can be purchased from the hardware store.

Required Precautions

- Read all safety precautions, and discuss them with your students.

- Safety goggles and a lab apron must be worn at all times.

- In case of a spill, use a dampened cloth or paper towels to mop up the spill. Then rinse the cloth in running water at the sink, wring it out until it is only damp and put it in the trash.

When a hot solid is immersed in a cool liquid, heat flows from the hot object to the cool liquid. In fact, the number of joules of energy lost by the hot solid (ΔQ_1) equals the number of joules of energy gained by the cool liquid (ΔQ_2).

$$- \Delta Q_1 = + \Delta Q_2$$

The quantity of heat that is lost or gained is a function of the object's mass and specific heat. Every material has a characteristic specific heat, which is the number of joules of energy needed to change the temperature of one gram of material one Celsius degree. Some specific heats are given in Table 1. If the mass, temperature change, and specific heat of a substance are known, the heat lost or gained can be calculated using this relationship: $\Delta Q = $ (mass)(specific heat)(Δt). The temperature change is defined as $\Delta t = t_{final} - t_{initial}$.

In this experiment, you will determine the temperature of a Bunsen burner flame by heating a sample of a known metal in the flame and then immersing the hot metal at temperature t_1 into a measured quantity of water at temperature t_2. As the heat flows from the hot metal to the cool water, the two materials approach an intermediate temperature t_3. These changes are shown graphically in Figure A.

$$- \Delta Q_1 = \Delta Q_2$$
$$- (m_1)(c_{p1})(t_3 - t_1). = (m_2)(c_{p2})(t_3 - t_2)$$

By solving the equation for t_1, the initial temperature of the hot metal can be found. The metal should be the same temperature as the Bunsen burner flame.

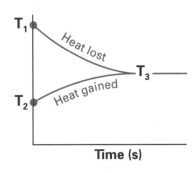

FIGURE A

SAFETY

Procedural Tips

• Suggests ways to attach the Nichrome wire to the metal object.

  Always wear safety goggles and a lab apron to provide protection for your eyes and clothing. If you get a chemical in your eyes, immediately flush the chemical out at the eyewash station while calling to your teacher. Know the location of the emergency lab shower and the eyewash station and the procedure for using them.

 When using a Bunsen burner, confine long hair and loose clothing. If your clothing catches on fire, WALK to the emergency lab shower and use it to put out the fire. Do not heat glassware that is broken, chipped, or cracked. Use tongs or a hot mitt to handle heated glassware and other equipment because hot glassware does not look hot.

 Never put broken glass in a regular waste container. Broken glass should be disposed of properly.

MATERIALS

• balance
• Bunsen burner
• foam cup
• iron ring

• metal object (Fe or Cu), 20–30 g
• Nichrome wire, 20 cm
• ring stand
• thermometer, nonmercury

PROCEDURE

• If students have difficulty with the equations in the Introduction, it may help to go through the calculations with a set of sample data. Some students may be interested in making the correction for the evaporated water discussed in Questions **2** and **3**. Suggest that they obtain the mass of the foam cup and water after the metal object is removed.

Disposal

• Save the metal objects for future labs.

CAUTION Remember that the metal sample will become very hot.

1. Obtain the mass of an empty foam cup to the nearest 0.01 g, and record the mass in the Data Table.

2. Fill the cup about two-thirds full with water. Obtain the mass of the cup and water and record it in the Data Table.

3. Read and record the temperature of the water to the nearest 0.1°C.

4. Obtain the mass of the metal object to the nearest 0.01 g and record it in the data table.

5. Use the Nichrome wire to attach the metal object to the iron ring clamped on the ring stand. The object should hang about 7 or 8 cm below the ring.

6. Adjust the Bunsen burner so that it produces a hot blue flame. Move the flame under the metal object, and heat it for about 5 minutes.

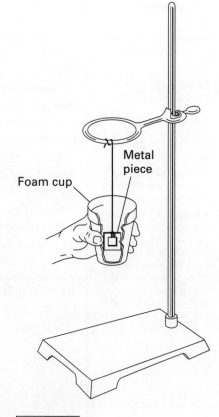

Metal piece

Foam cup

FIGURE B

7. Turn off the Bunsen burner and move it aside. SLOWLY lift the foam cup containing the water so that the hot metal becomes immersed in the water, as shown in Figure B. Hold the cup in this position for 1 min and then read and record the temperature of the water in the Data Table.

Data Table

Type of metal used	Ni	
Mass of metal	24.62	g
Specific heat of metal (from Table 1)	0.470	J/g·°C
Mass of empty foam cup	15.62	g
Mass of foam cup and water	225.32	g
Initial temperature of water	22.5	°C
Final temperature of water	33.2	°C

Table 1: Specific Heats of Materials

Substance	Specific heat (J/g·°C)
Copper	0.385
Iron	0.444
Manganese	0.481
Nickel	0.470
Platinum	0.131
Tungsten	0.134
Water	4.18

CALCULATIONS

1. **Organizing Data** Calculate the change in the temperature of the water.

   $\Delta t = 33.2°C - 22.5°C = 10.7°C$

2. **Organizing Data** What is the mass of the water that was heated by the metal object?

   mass of H_2O = 225.32 g − 15.62 g = 209.70 g

3. **Organizing Data** Calculate the heat gained by the water. (Hint: Refer to the Introduction.)

   mass × Δt × c_p = heat gained by the water
   209.70 g × 10.7°C × 4.18 J/g·°C = 9380 J

4. **Inferring Conclusions** Calculate the temperature of the hot metal object. (Hint: Refer to the Introduction. The heat lost by the metal equals the heat gained by the water.)

   24.62 g × 0.470 J/g°C × (t − 33.2°C) = 9380 J
   t = 844°C

5. Inferring Conclusions What is the temperature of the Bunsen burner flame?

844°C

1. Analyzing Information From the specific heats of metals listed in Table 1, which metal would raise the temperature of the water the most in this experiment? the least?

most: manganese; least: platinum

2. Evaluating Methods Error is present in every experiment. In this experiment, the small mass of water that was changed to steam when the hot sample was initially immersed in the water was neglected. Would this tend to make the calculated temperature lower or higher than the actual temperature of the flame. Explain your answer.

Lower. It takes energy to evaporate each gram of water. This is

heat that is neglected in the experiment.

3. Designing Experiments It takes about 2.26 kilojoules of energy to evaporate each gram of water. Suggest a way to include the evaporated water in your calculation.

Determine the mass of the foam cup and water after the metal

object has been removed. Find the mass of the missing water by

subtracting this mass from the mass of the foam cup and water

before the object was immersed. Multiply the missing mass by

2260 J/g, and add the result to the heat gained by the water.

4. Evaluating Methods Another source of error is the small amount of hot Nichrome wire that was heated and immersed in the water along with the metal object. Would this tend to make the calculated temperature lower or higher than the actual temperature of the flame? Explain your answer.

Higher. Heat from the small piece of hot Nichrome wire would

slightly increase the temperature of the water.

**GENERAL
CONCLUSIONS**

1. **Predicting Outcomes** If 100 g of boiling water at 100°C is poured into a 100 g platinum beaker at 0°C, an intermediate temperature will result. Knowing that the specific heat of water is about 32 times greater than the specific heat of platinum, approximate the intermediate temperature.

   97°C; For every Celsius degree lost by the water, the platinum

   gains 32 Celsius degrees.

2. **Applying Ideas** In the eighteenth century, clothes were pressed using the heat from a heavy piece of metal with a flat, smooth side. This solid metal "iron" had to be heated periodically on top of a wood- or coal-burning stove. Disregarding cost, which of the metals listed in Table 1 would be the best to use for this purpose? Explain your answer.

   Manganese. The metal with the largest specific heat absorbs the

   most heat energy from the stove.

EXPERIMENT **A17**

Heat of Solution

OBJECTIVES

Recommended time:
60 min

- **Measure** temperature changes in a calorimeter while a solid solute is dissolved.
- **Relate** temperature changes in a calorimeter to changes in heat energy.
- **Determine** the amount of heat released in joules for each gram of $Na_2S_2O_3 \cdot 5H_2O$ or $NaCH_3COO$ that dissolves.
- **Evaluate** the use of $Na_2S_2O_3 \cdot 5H_2O$ in a heating pad.

INTRODUCTION

Solution/Materials Preparation

1. Have chemical supply catalogs available as references for pricing $NaCH_3COO$ and $Na_2S_2O_3 \cdot 5H_2O$.

The solution in this heating pad is made by dissolving a salt in water. It is a supersaturated solution; that is, it holds more dissolved salt than is usually possible at a particular temperature. The solution is metastable, so disturbing it causes the ions to become ordered enough to crystallize out of solution, releasing heat energy. The same amount of heat energy is involved when crystals dissolve in water. The bombarding of the crystal lattice by water molecules causes the lattice to break apart. Then these free ions break the hydrogen bonds between water molecules. The water molecules surround the ions, attracted by their charge, and hydrate them. As these interactions take place, energy is released. The sum of the enthalpies of these processes is the heat of solution, ΔH_{sol}. If the hydration step, which involves bond formation, releases more heat than the bond-breaking step, the process is exothermic and ΔH_{sol} is negative. Otherwise, dissolving is endothermic and ΔH_{sol} is positive.

SAFETY

2. To save time, the complete thermometer/thermistor assemblies can be prepared in advance. Cut 15 cm lengths of wire. (Any gauge will do, provided it is easily bent.) Insert the length into one of the holes in the cardboard lid. At each end make a loop that has a diameter of about 1 cm. Bend the wire at the loops so that each loop is perpendicular to the rest of the wire.

Always wear safety goggles and a lab apron to protect your eyes and clothing. If you get a chemical in your eyes, immediately flush the chemical out at the eyewash station while calling to your teacher. Know the locations of the emergency lab shower and eyewash station and the procedure for using them.

Do not touch any chemicals. If you get a chemical on your skin or clothing, wash the chemical off at the sink while calling to your teacher. Carefully read the labels and follow the directions on all containers of chemicals that you use. Do not taste any chemicals or items used in the laboratory. Never return leftover chemicals to their original containers; take only small amounts to avoid wasting supplies.

MATERIALS

Procedural Tips

- Review identification and naming of ions, review the formation of ionic crystals, and review the summation of heat energies.

- Make sure students know that the quality of their technique is the focus of this experiment.

- 10–15 cm wire
- 100 mL graduated cylinder
- 600 mL beaker
- balance
- corrugated cardboard or lid for cup
- ice
- sharpened pencil
- plastic washtub
- plastic foam cup, small
- test tubes, small, 4

Probe option

- thermistor probe

Thermometer option

- thermometer, nonmercury

PROCEDURE

- Demonstrate how to carefully dump the salt from the test tube all at once. Remind students to pour the salt near the edge of the cup and not onto the thermometer to avoid false peak temperatures. The peak point of the temperature should be reached in 5 min or less.

- If the salt sticks to the side of the cup, a quick swirling motion can agitate enough water to wash the salt back into the cup.

- Demonstrate the use and the purpose of a stirrer.

Disposal

Set out two disposal containers. One is for sodium thiosulfate pentahydrate, $Na_2S_2O_3 \cdot 5H_2O$, and its solutions. The other is for sodium acetate, $NaC_2H_3O_2$, and its solutions. Allow the excess water to evaporate from the solutions, and store the solid salts in labeled reagent bottles for reuse. It will be necessary to pulverize the crystals into smaller chunks before reusing them.

1. Prepare a cold-water bath. Fill a small plastic washtub with ice. Fill a 600 mL beaker to three-fourths full with distilled water. Make a hole in the ice large enough for the beaker. Insert the beaker, and pack the ice around it up to the level of the water.

2. Prepare the calorimeter. Cut a square of corrugated cardboard slightly larger than the top of the plastic foam cup. Make a hole in the center of the cardboard piece with a pencil. Make a second hole in the center. Insert a piece of wire through the hole, and bend each end to make 1.0 cm loops, as shown in Figure A. Insert a thermometer or thermistor probe into the center hole, and set the entire assembly aside until step **8**.

3. On a piece of weighing paper, measure the mass of approximately 15 g of $Na_2S_2O_3 \cdot 5H_2O$. Pour the chemical into a small test tube. Record the mass of the $Na_2S_2O_3 \cdot 5H_2O$ in the Data Table.

Wire loop stirrer

Cardboard circle

Thermometer

FIGURE A

4. Label the tube Trial 1, and set it in a test-tube rack.

5. Repeat step **3** for the following
   a. 15 g of $Na_2S_2O_3 \cdot 5H_2O$; label the test tube *Trial 2*
   b. 15 g of $NaCH_3COO$; label the test tube *Trial 3*
   c. 15 g of $NaCH_3COO$; label the test tube *Trial 4*

6. Pour approximately 75 mL of the cold water from step **1** into a 100 mL graduated cylinder. Record the volume to the nearest 0.1 mL in the Data Table.

7. Pour the cold water from the graduated cylinder into the plastic foam cup.

8. Put the thermometer or thermistor assembly into the cup. The bulb of the thermometer should be completely covered by the water but must not touch the bottom of the cup. Record the temperature of the cold water to the nearest 1.0 °C in the Data Table.

9. Lift the cardboard slightly, and dump the entire contents of the Trial 1 test tube into the plastic foam cup. Gently move the stirring wire up and down inside the cup to disperse the heat. Allow the solid to dissolve completely. Stir continuously until the temperature of the water peaks.

10. Record the highest temperature reached by the water to the nearest 1.0 °C if using a thermometer or to the nearest 0.1 °C if using a thermistor probe.

11. Remove, rinse, and dry the thermometer or thermistor assembly. Pour the solution into the designated waste container. Rinse and dry the cup. Repeat steps **6** through **11** for Trial 2, Trial 3, and Trial 4.

Cleanup and Disposal

12. The $Na_2S_2O_3 \cdot 5H_2O$ and $NaCH_3COO$ and their solutions should each be placed in a designated disposal container. Remember to wash your hands thoroughly after cleaning up your lab area and equipment.

Data Table

Measurements	Trial 1— $Na_2S_2O_3 \cdot 5H_2O$	Trial 2— $Na_2S_2O_3 \cdot 5H_2O$	Trial 3— $NaCH_3COO$	Trial 4— $NaCH_3COO$
Mass of solute (g)	15.0	15.0	15.0	15.2
Volume of cold H_2O (mL)	75.0	75.0	75.0	75.0
Initial H_2O temp. (°C)	13.5	14.0	14.5	15.0
Final H_2O temp. (°C)	23.5	23.9	24.5	24.9

CALCULATIONS

1. **Organizing Data** Determine the change in temperature, Δt, of the cold water.

$\Delta t_1 = 23.5°C - 13.5°C = 10.0°C$

$\Delta t_2 = 23.9°C - 14°C = 9.9°C$

$\Delta t_3 = 24.5°C - 14.5°C = 10.0°C$

$\Delta t_4 = 24.9°C - 15.0°C = 9.9°C$

2. **Organizing Information** Calculate the heat energy in joules absorbed by the cold water for each trial, using the specific heat capacity equation. Assume the density of water is 1.00 g/mL and the specific heat capacity of water is 4.180 J/g·°C.

Heat absorbed $= m_{H_2O} \times c_{pH_2O} \times \Delta t$

$DH_2O = 1.00$ g/mL

75.0 mL $\times$ 1. g/mL $= 75.0$ g

Trial 1: 75.0 g $\times \dfrac{4.180\ J}{g \cdot °C} \times 10.0°C = 3.14 \times 10^3$ J

Trial 2: 75.0 g $\times \dfrac{4.180\ J}{g \cdot °C} \times 9.9°C = 3.10 \times 10^3$ J

Trial 3: 75.0 g $\times \dfrac{4.180\ J}{g \cdot °C} \times 10.0°C = 3.14 \times 10^3$ J

Trial 4: 75.0 g $\times \dfrac{4.180\ J}{g \cdot °C} \times 9.9°C = 3.10 \times 10^3$ J

3. **Organizing Data** Calculate the amount of heat energy released per gram of $Na_2S_2O_3 \cdot 5H_2O$ for Trial 1 and Trial 2.

Trial 1: $\dfrac{3140\ J}{15.0\ g} = 2.09 \times 10^2$ J/g $Na_2S_2O_3 \cdot 5H_2O$

Trial 2: $\dfrac{31\ 0\ J}{15.0\ g} = 2.07 \times 10^2$ J/g $Na_2S_2O_3 \cdot 5H_2O$

4. **Organizing Data** Calculate the amount of heat energy released per gram of $NaCH_3COO$ for Trial 3 and Trial 4.

Trial 3: $\dfrac{3140\ J}{15.0\ g} = 2.09 \times 10^2$ J/g $NaCH_3COO$

Trial 4: $\dfrac{3100\ J}{15.2\ g} = 2.04 \times 10^2$ J/g $NaCH_3COO$

5. **Analyzing Information** The accepted value for the heat of solution for $Na_2S_2O_3 \cdot 5H_2O$ is 200.0 J/g. Calculate your percent error.

$\dfrac{2.09 \times 10^2\ J/g - 2.000 \times 10^2\ J/g}{2.000 \times 10^2\ J/g} \times 100 = 4.5\%$ error

$\dfrac{2.0\ \times 10^2\ J/g - 2.000 \times 10^2\ J/g}{2.000 \times 10^2\ J/g} \times 100 = 3.5\%$ error

6. **Analyzing Information**   Calculate the average of Trial 1 and Trial 2, and the average of Trial 3 and Trial 4. Calculate the percent difference between each trial and the average.

average $= 2.08 \times 10^2$ J/g $Na_2S_2O_3 \cdot 5H_2O$
average $= 2.07 \times 10^2$ J/g $NaCH_3COO$

$$\frac{2.09 \times 10^2 \text{ J/g} - 2.08 \times 10^2 \text{ J/g}}{2.08 \times 10^2 \text{ J/g}} \times 100$$

$= 48\%$ difference for Trial 1
1.1% difference for Trial 2

$$\frac{2.09 \times 10^2 \text{ J/g} - 2.07 \times 10^2 \text{ J/g}}{2.07 \times 10^2 \text{ J/g}} \times 100$$

$= 0.97\%$ difference for Trial 3
1.5% difference for Trial 4

7. **Analyzing Information**   Locate a supplier, and find out the cost for $Na_2S_2O_3 \cdot 5H_2O$ and for $NaCH_3COO$. Calculate the cost per gram for each, and comment on which substance would be the most cost-effective primary ingredient of a heat pack.

cost of $Na_2S_2O_3 \cdot 5H_2O$: $0.0085/g

cost of $NaCH_3COO$: $0.0242/g

These values were calculated from a chemical supply catalog

that offered these chemicals in practical grade prices of

$8.50/kg and $12.10/500 g, respectively. Students' discussions

should indicate that thiosulfate appears more cost-effective

because nearly the same amount of energy per gram is provided

at a much cheaper cost.

QUESTIONS

1. **Evaluating Methods**   Why was a plastic foam cup instead of a beaker used as a reaction vessel?

The plastic foam is an insulator material, so very little heat is

transferred to the calorimeter or its surroundings. Glass is a

poorer insulator than plastic foam is; more heat is lost to the

calorimeter and the surroundings.

2. **Applying Ideas** When ionic substances dissolve in water, the crystal lattice is broken and the ions are immediately surrounded by water molecules in a process called hydration. Hydration is exothermic. Breaking the lattice absorbs heat energy. ΔH_{sol} is the net energy resulting from overcoming the lattice energy and releasing the energy of hydration. Draw a diagram relating ΔH_{lat}, ΔH_{hyd}, and ΔH_{sol}.

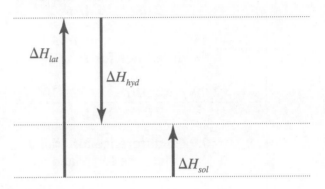

GENERAL CONCLUSIONS

1. **Analyzing Information** What is the source of the heat energy that is released by the $Na_2S_2O_3 \cdot 5H_2O$ and absorbed by the water?

   The ions lose potential energy as they become surrounded

   by water molecules. This energy is transferred to the

   water or the surroundings.

   _____

   _____

2. **Inferring Conclusions** What would be the ΔH of a reaction in which $NaCH_3COO$ crystallizes out of solution?

   A steady, rather than fluctuating, amount of heat is entering

   the surroundings. This reduces the risk of the temperature

   rising too high and the user being burned.

   _____

   _____

ChemFile

EXPERIMENT **A18**

Heat of Combustion

OBJECTIVES

Recommended time:
60 min

- **Measure** the change in temperature of a mass of water after heating it with a candle.

- **Use** the change in temperature, mass of the water, and specific heat of water to compute the energy released.

- **Determine** the energy released per mole of candle wax, or the heat of combustion.

- **Relate** heat of combustion to nutritional calories in food.

INTRODUCTION

Solution/Material Preparation

1. A 10 oz soup can is ideal for the calorimeter. Two holes for the stirring rod support are needed on opposite sides of the can and about 1 cm from the top.

The heat released by the complete combustion of one mole of a substance is called the *heat of combustion* of the substance. The *calorie* is sometimes used as the unit of energy for the heat of combustion because the nutritional Calorie content of foods is determined this way. The SI equivalent of one calorie is 4.184 joules. The calorie is defined as the quantity of heat required to raise the temperature of one gram of water one degree Celsius.

In this experiment, you will determine the heat of combustion of candle wax. A burning candle will heat a measured volume of water. You will observe the temperature change of the water. With this information you can calculate the heat released in the burning of the candle. The heat in calories is equal to the product of the mass of the water, its change in temperature, and the specific heat of water (1 cal/g • °C). With the molecular formula for candle wax and the mass of the wax consumed, the heat of combustion in calories per mole can be calculated.

SAFETY

2. Use a large fruit juice can (46 oz) for the chimney. Punch out at least four vents at the bottom with a can opener. Then cut out the bottom. Bend the sharp ends of the vents back and forth untiil they snap off. An alternate to punching out vents is to elevate the chimney on rubber stoppers.

  Always wear goggles and a lab apron to provide protection for your eyes and clothing. If you get a chemical in your eyes, immediately flush the chemical out at the eyewash station while calling to your teacher. Know the location of the emergency lab shower and the eyewash station and the procedure for using them.

 Call your teacher in the event of a spill. Spills should be cleaned up promptly, according to your teacher's directions.

 When you use a candle, confine long hair and loose clothing. If your clothing catches on fire, WALK to the emergency lab shower and use it to put out the fire. Do not heat glassware that is broken, chipped, or cracked. Use tongs or a hot mitt to handle heated glassware and other equipment because hot glassware does not look hot.

MATERIALS

Required Precautions

- Read all safety precautions, and discuss them with your students.

- 10 oz. tin can, open at one end
- 46 oz. tin can, open at both ends
- 100 mL graduated cylinder
- candle
- crucible tongs
- glass stirring rod
- ice cubes

- iron ring
- matches
- ring stand
- thermometer, nonmercury
- thermometer clamp
- tin can lid

PROCEDURE

- Safety goggles and a lab apron must be worn at all times.

- In case of a spill, use a dampened cloth or paper towels to mop up the spill. Then rinse the cloth in running water at the sink, wring it out until it is only damp and put it in the trash.

Procedural Tips

- Show students how to attach the candle to the tin can lid.

- Demonstrate how to set up the apparatus, especially how to position the calorimeter can so that it is the correct distance from the candle wick.

- If necessary, review specific heat and how it is used in calculations to convert temperature change to energy. Review the units of heat energy, calories, and joules and how one unit may be converted to the other.

Disposal

- Save candles for reuse or put into the trash.

1. Attach the candle to a tin can lid by lighting the candle and allowing a few drops of candle wax to drip on the lid before placing the candle on the lid. The tin can lid will be used to support the candle and collect any melted wax that runs down the side of the candle. The mass of the melted wax must be included in the mass of the candle after burning.

2. Obtain the mass of the candle and lid to the nearest 0.01 g, and record the mass in the Data Table.

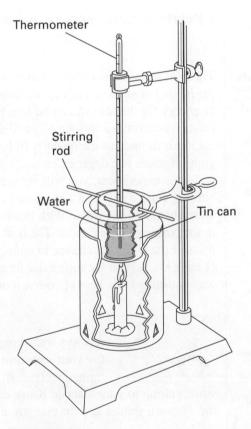

Thermometer

Stirring rod

Water

Tin can

FIGURE A

3. Insert the glass stirring rod through the two small holes on the sides of the small tin can. Support the can by the stirring rod on the iron ring and ring stand as shown in Figure A. With your candle in place, raise or lower the can to position it so that the bottom is approximately 5 cm above the top of the wick. The flame of the candle should just barely miss touching the bottom of the can. (You may wish to light the candle for a moment to see if you have the correct height.) Remove the can and stirring rod from the iron ring.

4. Place the large can over the candle. Make sure the air vents are at the bottom. Clamp a thermometer to the ring stand as shown in Figure A. Record the room temperature in the Data Table.

5. Fill the small can about half-full with tap water. Cool the water with ice until the temperature is 10–15°C below room temperature. Remove any remaining ice. Read and record the temperature of the water to the nearest 0.1°C.

6. Using crucible tongs to hold the match, light the candle. Immediately position the can of cold water on the iron ring and adjust the thermometer. While the candle heats the water, gently stir with a stirring rod. Make sure the lid catches all the drippings.

7. When the temperature is approximately the same number of degrees above room temperature as it was below, blow out the candle. Continue to stir the water and watch the temperature until the maximum temperature is reached. Record this temperature to the nearest 0.1°C.

8. Obtain the mass of the candle and lid to the nearest 0.01 g.

9. Use a graduated cylinder to measure the total volume of water in the can, and record the volume to the nearest 1 mL.

Cleanup and Disposal

10. Clean all apparatus and your lab station. Return equipment to its proper place. Wash your hands thoroughly before you leave the lab and after all work is finished.

Data Table

Initial mass of candle and base	52.90	g
Room temperature	22.0	°C
Initial temperature of water	12.0	°C
Final temperature of water	32.0	°C
Final mass of candle and base	51.94	g
Volume of water	338	mL

CALCULATIONS

1. **Organizing Data** Calculate the change in the temperature of the water. Record this result and all others in your calculations table.

   32.0°C − 12.0°C = 20.0°C

2. **Organizing Data** Determine the mass of the water. (Hint: The density of water is 1.00 g/mL.)

   338 mL × 1.00 g/mL = 338 g

3. **Organizing Data** Calculate the number of calories of heat absorbed by the can of water. (Hint: calories = mass of water × temperature change of water × specific heat of water in cal/g°C)

$$338 \text{ g} \times 20.0°\text{C} \times 1.00 \text{ cal/g}° \cdot \text{C} = 6760 \text{ cal}$$

4. **Organizing Data** Calculate the mass of the candle wax that burned.

$$52.90 \text{ g} - 51.94 \text{ g} = 0.96 \text{ g}$$

5. **Organizing Data** Calculate the heat released in the burning of one gram of candle wax.

$$\frac{6760 \text{ cal}}{0.96 \text{ g}} = 7100 \text{ cal/g}$$

6. **Inferring Conclusions** Your candle was probably made of a mixture of waxes, mostly paraffin waxes. Paraffin waxes are hydrocarbons with high molecular masses. Assume the molecular formula of the candle wax is $C_{36}H_{74}$. Calculate the heat of combustion of the candle wax in kcal/mol.

$$\frac{7100 \text{ cal}}{1 \text{ g}} \times \frac{506 \text{ g}}{1 \text{ mol}} \times \frac{1 \text{ kcal}}{1000 \text{ cal}} = 3600 \text{ kcal/mol}$$

7. **Inferring Conclusions** Convert the heat of combustion from Calculations item **6** to SI units of kJ/mol. (Hint: One calorie equals 4.184 J.)

$$\frac{3600 \text{ kcal}}{1 \text{ mol}} \times \frac{4.184 \text{ J}}{1 \text{ cal}} \times \frac{1000 \text{ cal}}{1 \text{ kcal}} \times \frac{1 \text{ kJ}}{1000 \text{ J}} = 15\,000 \text{ kJ/mol}$$

QUESTIONS

1. **Analyzing Methods** Explain the purpose of cooling the water to 10–15°C and then heating the water to the same number of degrees above room temperature.

   The purpose was to balance the heat gained from the environ-

   ment with the heat lost to the environment. When the water

   was below room temperature, heat was absorbed from the air.

   A similar amount of heat was released to the air when the water

   was heated above room temperature.

2. **Analyzing Methods** Was all the heat released in the burning of the candle absorbed by the water? If not, what happened to the missing heat?

   The heat of the flame radiated in all directions—into the air

   surrounding the candle, into the large can, and into the metal

   of the can containing the water.

GENERAL CONCLUSIONS

1. **Relating Ideas** How does the heat content of the candle wax compare with the heat content of the products of combustion? Draw an energy diagram to show the relationship.

   The heat content of the candle wax is greater than the heat con-

   tent of the products. The heat released in the burning is the dif-

   ference between the heat content of the reactants and the heat

   content of the products.

2. **Relating Ideas** A nutritional Calorie is a kilocalorie, or 1000 calories. The number of Calories that are in the foods you eat is determined using a precise bomb calorimeter, but the principle involved in the process is the same as that used in this experiment. Burning a peanut in a bomb calorimeter releases enough energy to raise the temperature of 500 grams of water by 10°C.

a. How many calories are contained in a peanut?

$$500 \text{ g} \times 10°C \times 1.0 \text{ cal/g}°\cdot C = 5000 \text{ calories}$$

b. How many nutritional Calories are contained in a peanut?

$$5000 \text{ cal} \times \frac{1 \text{ kcal}}{1000 \text{ cal}} = 5 \text{ kcal, or nutritional Calories}$$

Name _____

Date _____ Class _____

The Solubility Product Constant of Sodium Chloride

OBJECTIVES

Recommended time:
60 min

INTRODUCTION

Solution/Material Preparation

1. To make 100 mL of .1 M sodium acetate solution, add 0.82 g of CH_3COONa to enough water to make 100 mL of solution.

Required Precautions

• Safety goggles and a lab apron must be worn at all times.

• Tie back long hair and loose clothing when working in the lab.

• Read all safety cautions, and discuss them with your students.

SAFETY

Procedural Tips

• If necessary, review the technique of decanting. Caution students to allow the solid in the beaker containing the saturated NaCl solution to settle thoroughly before decanting.

• **Prepare** a saturated solution of NaCl.

• **Demonstrate** the principle of the common-ion effect in the precipitation of NaCl.

• **Determine** the solubility product constant for NaCl.

At equilibrium, a saturated solution contains the maximum amount of solute possible at a given temperature. This solute exists in equilibrium with an undissolved excess of the solute. For slightly soluble salts, the undissolved solid salt is in equilibrium with its ions in solution.

Because the concentration of a pure substance in the solid (or liquid) phase remains constant, the equilibrium constant expression may be expressed in a simplified form, called the *solubility product constant* expression. The solubility product constant of a substance is the product of the molar concentrations of its ions in a saturated solution, each raised to the power that is the coefficient of that ion in the chemical equation. For a simple salt composed of a $+1$ ion and a -1 ion, the equation for the equilibrium and the solubility product constant are as follows.

$$AB(s) \rightleftarrows A^+(aq) + B^-(aq)$$

$$K_{sp} = [A^+][B^-]$$

Sodium chloride is not considered a sparingly soluble compound, but its solubility is such that a K_{sp} may easily be determined experimentally. In this experiment, you will determine the value of the K_{sp} for NaCl and investigate how the addition of a common ion, the Na^+ ion, to a saturated solution of NaCl can initiate precipitation.

  Always wear safety goggles and a lab apron to protect your eyes and clothing. If you get a chemical in your eyes, immediately flush the chemical out at the eyewash station while calling to your teacher. Know the location of the emergency lab shower and the eyewash station and the procedure for using them.

 Do not touch any chemicals. If you get a chemical on your skin or clothing, wash the chemical off at the sink while calling to your teacher. Make sure you carefully read the labels and follow the precautions on all containers of chemicals that you use. If there are no precautions stated on the label, ask your teacher what precautions you should follow. Never return leftovers to their original containers.

- Discuss the form of the solubility product constant expression.

- Make sure students understand that solubility equilibrium involves the equilibrium between the processes of dissolving and precipitating, and therefore, some solid will always be present when equilibrium is established.

 Call your teacher in the event of a spill. Spills should be cleaned up promptly, according to your teacher's directions.

 When you use a Bunsen burner, confine long hair and loose clothing. If your clothing catches on fire, WALK to the emergency lab shower, and use it to put out the fire. Do not heat glassware that is broken, chipped, or cracked. Use tongs or a hot mitt to handle heated glassware and other equipment because hot glassware does not always look hot.

 Never put broken glass in a regular waste container. Broken glass should be disposed of properly.

MATERIALS

Disposal

- Combine all solutions and pour down the drain. Combine all solids and put into the trash.

- 0.1 M CH₃COONa solution
- 10 mL graduated cylinder
- 25 × 100 mm, test tubes, 2
- 50 mL graduated cylinder
- 150 mL beakers, 2
- balance
- Bunsen burner and related equipment
- evaporating dish

- glass stirring rod
- iron ring
- CH₃COONa
- NaCl
- ring stand
- sparker
- tongs
- wire gauze, ceramic-centered

PROCEDURE

1. To prepare a saturated solution of sodium chloride, measure 25 mL of water in the graduated cylinder, and pour it into a beaker. Add 10 g of NaCl and stir constantly for a minute. There should be some undissolved solid on the bottom of the beaker. Allow the solid to settle, and then decant the clear, saturated solution into a clean beaker.

2. Determine the mass of the evaporating dish to the nearest 0.01 g, and record the mass in the Data Table.

3. Use a graduated cylinder to measure approximately 10 mL of the salt solution. Record the volume to the nearest 0.1 mL. Pour this portion of the saturated solution into the evaporating dish.

4. Assemble the ring stand, ring, wire gauze, and Bunsen burner. Clamp the ring assembly several centimeters above the flame so that you can heat gently and avoid spattering. Place the evaporating dish on the wire gauze, and heat the solution to evaporate it. When the solution is close to dryness, reduce the burner to a very low flame, and continue heating for 10 minutes.

5. When heating is complete, remove the evaporating dish with tongs, and place it on the base of the ring stand to cool. Then determine the mass of the evaporating dish and solid residue to the nearest 0.01 g. Record the mass in the Data Table.

6. This step covers the addition of a common ion to a saturated solution of salt.
   a. Measure 5 mL of the saturated NaCl solution, and pour it into a clean test tube. Measure 1 g of sodium acetate, CH₃COONa, and add it to the solution. Stir to dissolve as much sodium acetate as possible. Record your observations in the Data Table.

b. Measure another 5 mL of the saturated NaCl solution into a test tube, and add 5 mL of 2 M CH_3COONa solution. Stir thoroughly, and then record your observations. Remember that to be certain a solution is saturated, some solid must remain undissolved after a period of stirring.

Cleanup and Disposal

7. Clean all apparatus and your lab station. Return equipment to its proper place. Dispose of chemicals and solutions in the containers designated by your teacher. Do not pour any chemicals down the drain or in the trash unless your teacher directs you to do so. Make sure to shut off the gas valve completely before leaving the laboratory. Wash your hands thoroughly after all work is finished and before you leave the lab.

Data Table

Mass of empty evaporating dish	41.36	g
Volume of saturated NaCl solution	10.0	mL
Mass of evaporating dish and NaCl	44.50	g

Observations step 6a:	A white precipitate forms.
Observations step 6b:	The solution remains clear.

CALCULATIONS

1. Organizing Data Determine the mass of dry NaCl residue.

Mass of NaCl = 44.50 g − 41.36 g = 3.14 g

2. Organizing Data Calculate the moles of dry NaCl.

Moles of NaCl = $3.14 \text{ g} \times \dfrac{1 \text{ mol}}{58.44 \text{ g}} = 0.0537 \text{ mol}$

3. Organizing Data Calculate the concentration of the original 10 mL of NaCl solution before drying.

Molarity of NaCl = $\dfrac{0.0537 \text{ mol}}{10.0 \text{ mL}} \times \dfrac{1000 \text{ mL}}{1 \text{ L}} = 5.37 \text{ M}$

4. Organizing Information Write the equilibrium equation for the saturated solution and the solubility product constant (K_{sp}) expression for the system.

$NaCl(s) \rightleftarrows Na^+(aq) + Cl^-(aq)$

$K_{sp} = [Na^+][Cl^-]$

5. Organizing Conclusions Determine the concentrations of Na^+ ions and Cl^- ions using the balanced equation in Calculations item **4** and the molarity of the saturated NaCl solution. (Hint: One mole of NaCl dissociates into one mole of Na^+ ions and one mole of Cl^- ions.)

$[NaCl] = [Na^+] = [Cl^-] = 5.37$ M

6. Organizing Conclusions Calculate the K_{sp} for sodium chloride.

$K_{sp} = [Na^+][Cl^-] = [5.37][5.37] = 28.8$

QUESTIONS

1. Analyzing Methods Explain your observations for step **6a.**

The additional Na^+ ions from the sodium acetate caused the product of the concentrations of the Na^+ ion and Cl^- ion to exceed the solubility product constant of NaCl, resulting in the precipitation of solid NaCl.

2. Resolving Discrepancies Explain any differences in the observations you made for steps **6a** and **6b.**

The addition of 2 M CH_3COONa solution did not produce a precipitate because the concentration of Na^+ ions and Cl^- ions to exceed the K_{sp} for NaCl. This is because the water added along with the CH_3COONa in the 2 M solution decreased the concentration of the original ions present.

GENERAL CONCLUSIONS

1. Predicting Outcomes Explain what would have happened if you had added 1 g of KCl to 5 mL of saturated sodium chloride solution in step **6a.** If your teacher approves, test your prediction.

A precipitate would have formed because the increased concentration of Cl^- ions would have caused the product of the concentrations of Na^+ ions and Cl^- ions to exceed the K_{sp} for NaCl.

EXPERIMENT A20

Buffering Capacity

OBJECTIVES

Recommended time:
2 lab periods

Solution/Material Preparation

1. Wear safety goggles, a face shield, impermeable gloves, and a lab apron when you prepare the HCl and NaOH. When preparing the HCl,

- **Demonstrate** proficiency in measuring pH and performing a titration.

- **Relate** the ability of a solution to absorb acid or base while maintaining its pH to its buffering capacity.

- **Determine** which ratio of acetic acid to sodium acetate provides the most efficient buffer.

- **Graph** the curve for the titration of the acetic acid and sodium acetate buffer system.

- **Interpret** the shape of the buffer titration curve.

- **Calculate** the buffering capacity for the ratio you determined.

INTRODUCTION

work in a hood known to be in operating condition and have another person stand by to call for help in case of an emergency. Be sure you are within a 30 s walk of a safety shower and eyewash station known to be in good operating condition.

The effectiveness of any buffer system depends on the ratio of weak acid to weak base in the solution. For every ratio, there is a limit to how many hydronium ions or hydroxide ions any buffer system can absorb. Eventually, the pH rises or drops substantially because the limit has been exceeded. The number of hydronium ions or hydroxide ions that can be added to a buffer before the pH changes is called the *buffering capacity* of the system.

In this experiment, you will investigate the acetic acid and sodium acetate buffer system and determine the ratio of acetic acid to sodium acetate that provides the best buffer. You will then titrate a buffer solution having this ratio and determine its buffering capacity.

SAFETY

2. To prepare 0.10 M CH_3COOH, observe the required safety precautions. Slowly and with stirring, add 5.7 mL of glacial acetic acid to enough distilled water to make 1 L of solution.

3. To prepare 0.10 M CH_3COONa, dissolve 13.6 g of $CH_3COONa \cdot 3H_2O$ in enough distilled water to make 1 L of solution.

  Always wear safety goggles and a lab apron to protect your eyes and clothing. If you get a chemical in your eyes, immediately flush the chemical out at the eyewash station while calling to your teacher. Know the location of the emergency lab shower and the eyewash station and the procedure for using them.

 Do not touch any chemicals used in the laboratory. If you get a chemical on your skin or clothing, wash the chemical off at the sink while calling to your teacher. Make sure you carefully read the labels and follow the precautions on all containers of chemicals that you use. If there are no precautions stated on the label, ask your teacher what precautions to follow. Never return leftovers to their original containers; take only small amounts to avoid wasting supplies.

 Call your teacher in the event of a spill. Spills should be cleaned up promptly, according to your teacher's directions.

 Never put broken glass into a regular waste container. Broken glass should be disposed of separately according to your teacher's instructions.

MATERIALS

- 0.10 M CH_3COOH
- 0.10 M CH_3COONa
- 1.0 M HCl in a dropper bottle
- 1.0 M NaOH in a dropper bottle
- 10 mL graduated cylinder

- 50 mL beakers, 2
- 400 mL beaker
- distilled water
- pH meter or pH paper

PROCEDURE

4. To prepare 1.00 M HCl, observe the required safety precautions. Add 82.6 mL of concentrated HCl to enough distilled water to make 1 L of solution. Add the acid slowly, and stop to stir it in order to avoid overheating.

5. To prepare 1.00 M NaOH, add 40.0 g of NaOH to enough distilled water to make 1 L of solution. Add a few pellets at a time, and stop to stir the solution in order to avoid overheating.

6. Check the pH of your distilled water; it should be approximately 7. If it is low, boil the water shortly before the lab to drive off dissolved carbon dioxide. Store the boiled water in containers with tight-fitting lids.

7. A pH meter will provide far more precise data than even narrow-range pH paper.

Required Precautions

• Safety goggles and a lab apron must be worn at all times.

• Read all safety cautions, and discuss them with your students.

• In case of an acid or base spill, first dilute with water. Then, mop up the spill with wet cloths or a wet cloth mop while wearing disposable plastic gloves. Designate separate cloths or mops for acid and base spills.

1. Calibrate the pH meter using a standard buffer of known pH. This step is not necessary if you are using pH paper.

2. Label the two 50 mL beakers *Solution* and *Solution + HCl*. Label the 400 mL beaker *Waste*.

Determining Best Ratio for Buffer

3. Measure 10.0 mL of sodium acetate in a 10 mL graduated cylinder, and pour it into the 50 mL beaker labeled *Solution*. Measure 10.0 mL of distilled water, and add it to the beaker, swirling the mixture gently to mix thoroughly. This is Solution 1, as shown in Figure A.

4. Measure the initial pH of Solution 1, and record it in Data Table 1 under *Original pH*.

5. Pour about half of Solution 1 into the beaker labeled *Solution + HCl*. Add one drop of 1.0 M HCl as shown in Figure A, and swirl the beaker gently. Measure and record the pH in Data Table 1.

6. Continue adding HCl drop by drop. After each drop, record the pH. Stop when 5 drops have been added.

7. Repeat steps **3–6** with the remainder of Solution 1 that is left in the beaker labeled *Solution*, but this time add drops of 1.0 M NaOH. Measure and record the pH in Data Table 1.

8. Empty both 50 mL beakers into the *Waste* beaker. Clean both beakers and the graduated cylinder. Rinse them several times with distilled water.

9. Repeat steps **4–8** for Solutions 2–7. For best results, prepare solutions one at a time.
- Solution 2 should contain 8.0 mL of CH_3COONa, 2.0 mL of CH_3COOH, and 10.0 mL of distilled water.
- Solution 3 should contain 6.0 mL of CH_3COONa, 4.0 mL of CH_3COOH, and 10.0 mL of distilled water.
- Solution 4 should contain 5.0 mL of CH_3COONa, 5.0 mL of CH_3COOH, and 10.0 mL of distilled water.
- Solution 5 should contain 4.0 mL of CH_3COONa, 6.0 mL of CH_3COOH, and 10.0 mL of distilled water.
- Solution 6 should contain 2.0 mL of CH_3COONa, 8.0 mL of CH_3COOH, and 10.0 mL of distilled water.
- Solution 7 should contain 10.0 mL of CH_3COOH and 10.0 mL of distilled water.

10. Clean the beakers and graduated cylinder. Rinse each with distilled water before proceeding to step **11.**

Procedural Tips

• Demonstrate the proper titration method and the procedure for calibrating the pH meter, if used.
• Review the discussion from the Introduction that describes how buffers are able to maintain constant pH even when acid or base is added. Be sure to link this discussion with the concepts of equilibrium.

Determining Buffering Capacity

11. Examine the pH measurements in Data Table 1. Calculate the difference between the pH obtained after adding 5 drops of HCl and the pH obtained after adding 5 drops of NaOH to each solution. Which ratio of weak acid to weak base shows the smallest change in pH whether acid or base is added?

12. In the beaker labeled *Solution + HCl,* prepare another batch of the solution you identified in step **11** as the best buffer. Add 1.0 M HCl one drop at a time. Measure and record the pH after each drop in Data Table 2. Stop adding HCl when the pH drops suddenly, indicating that the solution has lost its buffering capacity. Empty the solution into the *Waste* beaker.

13. In the beaker labeled *Solution,* prepare another batch of the solution you identified in step **11** as the best buffer. Add 1.0 M NaOH one drop at a time. Measure and record the pH after each drop in Data Table 2. Stop adding NaOH when the pH rises suddenly, indicating that the solution has lost its buffering capacity. Empty the solution into the *Waste* beaker.

Cleanup and Disposal

14. Pour the solution in the *Waste* beaker into the container designated by your teacher. Place any leftover acid solutions in a designated container. Leftover basic solutions should be placed in a different designated container. Thoroughly clean the area. Wash your hands after all work is completed and before you leave the lab.

Data Table 1—*Best Ratio*

Solution	Original pH	pH after 1 drop	pH after 2 drops	pH after 3 drops	pH after 4 drops	pH after 5 drops
1 HCl	7.2	6.4	6.2	6.1		
1 NaOH	7.2	11.2	11.5	11.4		
2 HCl	5.9	5.8	5.7	5.6		
2 NaOH	5.9	6.0	6.2	6.4		
3 HCl	5.5	5.4	5.3	5.3		
3 NaOH	5.4	5.6	5.7	5.8		
4 HCl	5.4	5.4	5.3	5.2		
4 NaOH	5.4	5.4	5.5	5.6		
5 HCl	5.2	5.2	5.1	4.9		
5 NaOH	5.2	5.3	5.4	5.5		
6 HCl	4.7	4.6	4.5	4.3		
6 NaOH	4.7	4.8	4.9	5.5		
7 HCl	4.4	4.1	3.9	3.7		
7 NaOH	4.4	4.6	4.9	5.0		

Data Table 2—*Buffering Capacity*

Solution	HCl	NaOH	Solution	HCl	NaOH
Original pH	4.9	4.9	After 11 drops	3.1	5.9
After 1 drop	4.9	4.9	After 12 drops	2.9	6.2
After 2 drops	4.9	5.1	After 13 drops	2.8	6.5
After 3 drops	4.9	5.2	After 14 drops	2.7	7.7
After 4 drops	4.6	5.3	After 15 drops	2.6	10.4
After 5 drops	4.5	5.3	After 16 drops	2.5	10.8
After 6 drops	4.4	5.3	After 17 drops	2.4	11.0
After 7 drops	4.2	5.5	After 18 drops	2.3	11.3
After 8 drops	3.9	5.6	After 19 drops	2.2	11.3
After 9 drops	3.6	5.7	After 20 drops	2.1	11.9
After 10 drops	3.3	5.8			

CALCULATIONS

Disposal

• Set out three disposal containers: one for unused acid solutions, one for unused base solutions, and one for partially neutralized substances and the contents of the *Waste* beaker. One at a time, slowly combine the solutions while stirring. Adjust the pH of the final waste liquid with 1.0 M acid or base until the pH is between 5 and 9. Pour the neutralized liquid down the drain.

1. Organizing Data Draw a graph below and plot your pH data from the two titrations. To the left of the vertical pH line, plot the increasing numbers of drops of HCl. To the right of the line, plot the increasing numbers of drops of NaOH.

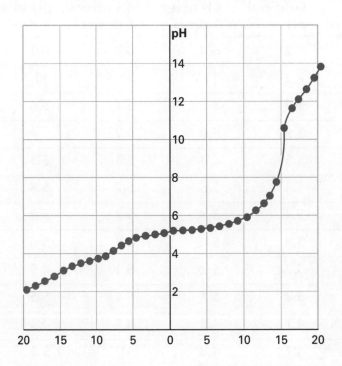

QUESTIONS

1. **Applying Ideas** If you needed a solution that had to maintain the same pH with additions of acid only, which solution would you pick? If you needed a solution that had to maintain the same pH with additions of base only, which solution would you pick?

Solution 3 would maintain a good pH when acids were added,

and Solution 5 would maintain a good pH when bases were

added.

2. **Analyzing Data** Using your data, explain why neither of the solutions you chose in Question **1** was the buffer you chose in step **11.**

Of all the solutions, solution 4 kept an average pH closest to the

original pH of 4.9 whether 5 drops of acid or base were added.

GENERAL CONCLUSIONS

1. **Inferring Conclusions** Which ratio of acetic acid to sodium acetate provides the best buffering system? Use your data to support your choice.

Solution 4 has a 1:1 ratio of acid to base. The change in pH for

this ratio is the smallest of the seven ratios.

2. **Interpreting Graphics** Examine the way that the slope of your titration curve changes. In what pH range does the slope change the least? In what pH range would the acetic acid and sodium acetate buffer be most suitable? Could you use this buffer system to run a reaction that requires a pH of approximately 5?

For the sample data, the slope changes the least in the pH

range from 4.0 to 6.0. This is the most useful range for this buffer.

It is appropriate for maintaining a pH of approximately 5.

3. **Applying Ideas** For the buffer you chose, how many drops of acid or base was it able to absorb before the pH changed?

For the sample data, 7 drops of acid were added before the pH

went below 4.0; 11 drops of base were added before the pH went

above 6.0.

4. **Applying Ideas** Write the balanced equilibrium equation for the dissociation of acetic acid. Also write the equilibrium expression. Use a chemical handbook to find the numerical value for the acid-ionization constant, K_a. Substitute the numerical values for K_a and the initial concentrations of CH_3COOH and CH_3COO into the equilibrium expression. Solve the equation for the H_3O^+, and convert your answer to pH.

$$CH_3COOH(aq) + H_2O(l) \rightarrow CH_3COO^-(aq) + H_3O^+(aq)$$

$$K_a = \frac{[CH_3COO][H_3O^+]}{[CH_3COOH]} = 1.75 \times 10^{-5}$$

$$K_a = \frac{(0.10)[H_3O^+]}{(0.10)} \qquad [H_3O^+] = 1.8 \times 10^{-5} \qquad pH = 4.7$$

5. **Applying Ideas** Using the value of the acid-ionization constant, K_a, explain why the assumption was made that the equilibrium concentrations were the same as the initial concentrations for CH_3COOH and CH_3COO^-.

Because the value of the constant is a small number, relatively small

amounts of the products are formed. Thus, the equilibrium concen-

tration of the original reactants will be very close to the initial con-

centrations.

6. **Applying Conclusions** Chemists officially define buffering capacity as the number of moles of H_3O^+ or OH^- needed to cause 1.00 L of the buffered solution to undergo a 1.00-unit change in pH. Calculate the buffering capacity for your solution. (Hint: in order to figure out the number of moles of H_3O^+ or OH^-, you will need to determine the volume of a drop. This can be done by measuring the volume of 50 drops and dividing to determine the volume of an individual drop.)

Student results for calibrating the dropper will vary. Sample analy-

sis: 40 drops/mL

For 1.00-unit change in pH, 8 drops of HCl or 11 drops of NaOH

are needed.

$$1000 \text{ mL buffer} \times \frac{8 \text{ drops HCl}}{20 \text{ mL buffer}} \times \frac{1.0 \text{ mL acid}}{40 \text{ drops HCl}} \times \frac{1 \text{ L}}{1000 \text{ mL}}$$

$$\times \frac{1.0 \text{ mol HCl}}{1 \text{ L}} = 0.010 \text{ mol HCl}$$

$$1000 \text{ mL buffer} \times \frac{11 \text{ drops NaOH}}{20 \text{ mL buffer}} \times \frac{1.0 \text{ mL acid}}{40 \text{ drops NaOH}} \times \frac{1 \text{ L}}{1000 \text{ mL}}$$

$$\times \frac{1.0 \text{ mol NaOH}}{1 \text{ L}} = 0.010 \text{ mol HCl}$$

EXPERIMENT **A21**

Oxidation-Reduction Reactions

OBJECTIVES

Recommended time:
60 min

- **Observe** oxidation-reduction reactions between metals.
- **Describe** typical oxidation-reduction reactions.
- **Determine** relative strengths of some oxidizing and reducing agents.

INTRODUCTION

Solution/Material Preparation

1. Wear safety goggles, a face shield, impermeable gloves, and a lab apron when you prepare the iron(III) chloride solution containing HCl and the 1.0 M H_2SO_4. Work in a hood known to be in operating condition and have another person stand by to call for help in

A substance that loses electrons during a chemical reaction is said to be oxidized. A substance that gains electrons is said to be reduced. If one reactant gains electrons, another must lose an equal number. Thus, oxidation and reduction must occur simultaneously and to a comparable degree.

The stronger the tendency for a species to take electrons, the greater is its strength as an oxidizing agent and the more easily it is reduced. The stronger the tendency for a species to give up electrons, the greater is its strength as a reducing agent and the more readily it is oxidized. The silver ion, Ag^+, has a strong tendency to acquire an electron to form the silver atom, Ag. Thus, the Ag^+ ion is a strong oxidizing agent. In this experiment, you will determine the relative strengths of some metals as reducing agents and the relative strengths of their ions as oxidizing agents.

SAFETY

case of an emergency. Be sure you are within a 30 s walk of a safety shower and eyewash station known to be in good operating condition.

2. To prepare 1 L of 0.1 M copper(II) nitrate, dissolve 24 g of $Cu(NO_3)_2 \cdot 6H_2O$ in enough distilled water to make 1 L of solution.

3. To prepare 1 L of 0.1 M zinc nitrate, dissolve 30 g $Zn(NO_3)_2$ in enough distilled water to make 1 L of solution.

4. To prepare 1 L of 0.1 M magnesium nitrate, dissolve 18 g of $Mg(NO_3)_2 \cdot 2H_2O$ in enough distilled water to make 1 L of solution.

  Always wear safety goggles and a lab apron to protect your eyes and clothing. If you get a chemical in your eyes, immediately flush the chemical out at the eyewash station while calling to your teacher. Know the location of the emergency lab shower and the eyewash station and the procedure for using them.

 Do not touch any chemicals used in the laboratory. If you get a chemical on your skin or clothing, wash the chemical off at the sink while calling to your teacher. Make sure you carefully read the labels and follow the precautions on all containers of chemicals that you use. If there are no precautions stated on the label, ask your teacher what precautions to follow. Never return leftovers to their original containers; take only small amounts to avoid wasting supplies.

 Call your teacher in the event of a spill. Spills should be cleaned up promptly, according to your teacher's directions.

 When you insert glass tubing into stoppers, lubricate the glass with water or glycerin and protect your hands and fingers. Wear leather gloves or place folded cloth pads between both of your hands and the

ChemFile   EXPERIMENT A21 **125**

HRW material copyrighted under notice appearing earlier in this work.

glass tubing. Then *gently* push the tubing into the stopper hole. In the same way, protect your hands and fingers when removing glass tubing from stoppers and from rubber or plastic tubing.

 Never put broken glass into a regular waste container. Broken glass should be disposed of separately according to your teacher's instructions.

MATERIALS

5. To prepare 500 mL of 0.1 M iron(III) chloride, observe the required safety precautions. Slowly and with stirring, add 20 mL of concentrated HCl to about 800 mL of distilled water. Then add 14 g of FeCl₃•6H₂O and dilute the solution to a final volume of 500 mL.

6. To prepare 1 L of 0.1 M potassium permanganate, dissolve 16 g of KMnO₄ in enough distilled water to make 1 L of solution.

- 0.1 M $Cu(NO_3)_2$
- 0.1 M $FeCl_3$
- 0.1 M $Mg(NO_3)_2$
- 0.1 M $KMnO_4$
- 0.1 M $Zn(NO_3)_2$
- 01. M H_2SO_4
- balance
- copper strips, 1 cm × 5 cm, 2
- $FeSO_4$
- forceps
- magnesium ribbon, 10 cm
- medicine dropper
- sandpaper
- $SnCl_2$
- test tubes, 18 mm × 150 mm, 7
- zinc strips, 1 cm × 5 cm, 2

PROCEDURE

7. To prepare 1 L of 1.0 M sulfuric acid, observe the required safety precautions. Slowly and with stirring, add 56 mL of concentrated H_2SO_4 to enough distilled water to make 1 L of solution.

Required Precautions

- Read all safety precautions, and discuss them with your students.

- Safety goggles and a lab apron must be worn at all times.

- In case of an acid or base spill, first dilute with water. Then, mop up the spill with wet cloths or a wet cloth mop while wearing disposable plastic gloves.

1. Sand a strip of zinc until all zinc oxide has been removed and the surface is shiny. Place the zinc in a test tube containing approximately 5 mL of $Cu(NO_3)_2$ solution. Place another strip of zinc into approximately 5 mL of $Mg(NO_3)_2$ solution. After a few minutes, examine both pieces of metal. Record your observations in the appropriate space in the Data Table.

2. Place a strip of shiny magnesium into approximately 5 mL of $Zn(NO_3)_2$ solution. Place another strip of shiny magnesium into approximately 5 mL of $Cu(NO_3)_2$ solution. After a few minutes, examine both pieces of metal and record your observations in the Data Table.

3. Place a strip of copper into approximately 5 mL of $Zn(NO_3)_2$ solution. Place another strip of copper into 5 mL of $Mg(NO_3)_2$. After a few minutes, examine both strips of metal, and record your observations in the Data Table.

4. Pour 1 to 2 mL of $FeCl_3$ solution into a test tube, and record its color. Completely dissolve a few crystals of $SnCl_2$ in the solution and record the color change.

5. Completely dissolve 0.5 g of $FeSO_4$ in 25 mL of water. Pour 5 mL of this solution into a large test tube. Add 10 drops of 1.0 M H_2SO_4. Then add a drop of 0.1 M $KMnO_4$ solution and mix thoroughly. Continue to add $KMnO_4$ dropwise and mix until two color changes have occurred. Record your results in the Data Table.

Procedural Tips

• Encourage students to think about the reactions in terms of changing oxidation states.
• Make sure they understand that when a metal dissolves in water, it has become an ion and therefore has lost electrons (been oxidized). Conversely, if a metallic ion gains electrons (is reduced), it is no longer soluble in water and will precipitate.

Cleanup and Disposal

6. Clean all apparatus and your lab station. Return equipment to its proper place. Dispose of chemicals and solutions in the containers designated by your teacher. Do not pour any chemicals down the drain or in the trash unless your teacher directs you to do so. Wash your hands thoroughly after all work is finished and before you leave the lab.

Data Table

	Cu	Zn	Mg
$Cu(NO_3)_2$	NR	Zn dissolves, Cu ppt	Mg dissolves, Cu ppt
$Zn(NO_3)_2$	NR	NR	Mg dissolves, Zn ppt
$Mg(NO_3)_2$	NR	NR	NR
$FeCl_3 + SnCl_2$	orange to light green color change		
$KMnO_4 + FeSO_4$	green to orange to deep purple color change		

QUESTIONS

• Emphasize the point that strong oxidizing agents are readily reduced and strong reducing agents are readily oxidized.

Disposal

Save the metal strips for reuse unless they are too corroded, in which case you should put them into the trash. Combine all solutions. If the combined solution is purple or even slightly pink in color, add 1.0 M $Na_2S_2O_3$ until decolorized. Then adjust the pH to between 5 and 9, dilute tenfold, and pour down the drain.

1. Organizing Data Which metal was oxidized by two other ions?

magnesium

2. Organizing Data Which metal was oxidized by only one other ion?

zinc

3. Organizing Data Which metal was not oxidized by any of the ions?

copper

4. Analyzing Results Arrange the three metals in order of their relative strengths as reducing agents, placing the strongest first.

Mg^0, Zn^0, Cu^0

5. Analyzing Results Arrange the three metallic ions in order of their relative strengths as oxidizing agents, placing the strongest first. Write the reduction half-reaction for each reaction.

1. $Cu^{2+} + 2e^- \rightarrow Cu^0$

2. $Zn^{2+} + 2e^- \rightarrow Zn^0$

3. $Mg^{2+} + 2e^- \rightarrow Mg^0$

6. Inferring Results Copper is oxidized in the presence of silver ions. The net ionic reaction is the following: $Cu + 2Ag^+ \rightarrow Cu^{2+} + 2Ag$
Write net ionic equations for (a) the reaction of copper and zinc, (b) the reaction of zinc and magnesium, and (c) the reaction of copper and magnesium. (Hint: From your answers to Questions **4** and **5,** determine which metal is oxidized and which ion is reduced.)

a. $Cu^{2+} + Zn \rightarrow Cu^0 + Zn^{2+}$

b. $Zn^{2+} + Mg^0 \rightarrow Zn^0 + Mg^{2+}$

c. $Cu^{2+} + Mg^0 \rightarrow Cu^0 + Mg^{2+}$

7. **Analyzing Results**   In step **4**, the Fe^{3+} ion was reduced to the Fe^{2+} ion.
   a. What was the reducing agent? **b.** What change did the Sn^{2+} ion undergo?
   c. Write the net ionic equation for the overall reaction:
   $2Fe^{3+} + 6Cl^- + Sn^{2+} + 2Cl^- \rightarrow 2Fe^{2+} + 4Cl^- + Sn^{4+} + 4Cl^-$

   a. Sn^{2+} ion,

   b. Sn^{2+} was oxidized to Sn^{4+}.

   c. $2Fe^{3+} + Sn^{2+} \rightarrow 2Fe^{2+} + Sn^{4+}$

8. **Analyzing Results**   The permanganate ion, MnO_4^-, which is purple in
   color, is a strong oxidizing agent. The manganese(II) ion, Mn^{2+}, is practically
   colorless. What occurred during the addition of the permanganate ions to the
   Fe^{2+} ions?

   The almost colorless (pale green) solution of Fe^{2+} changes to the

   yellow-brown of Fe^{3+}.

**GENERAL
CONCLUSIONS**

1. **Inferring Conclusions**   An unknown metal, X, was found to react with
   magnesium(II) ions and with zinc ions, but not with copper ions. Insert it in
   the series in Question **4** and Question **5**. What will be the new series?

   Question 4: Mg, Zn, Cu

   Question 5: Cu, Zn, Mg

2. **Predicting Outcomes**   Describe what would happen to the metals if
   iron nails were used to secure sheets of copper to a roof.

   Iron is a more active metal than copper. Thus, the nails will

   corrode, causing the sheets of copper to become loose.

Name _____

Date _____ Class _____

HOLT
ChemFile
LAB PROGRAM

EXPERIMENT **A22**

Cathodic Protection: Factors Affecting the Corrosion of Iron

OBJECTIVES

Recommended time:
60 min with about 20 min the day before and 10 min the day after the lab

- **Identify** factors that affect the rate of corrosion of iron.
- **Use** a control for comparison.
- **Recognize** corrosion of iron as the result of oxidation-reduction.
- **Describe** ways of reducing corrosion.

INTRODUCTION

Solution/Material Preparation

1. To prepare 500 mL of 0.1 M potassium hexacyanoferrate(III), dissolve 17 g of $K_3Fe(CN)_6$ in enough water to make 500 mL of solution.

2. To prepare 100 mL of phenolphthalein indicator, dissolve 1.0 g of phenolphthalein in 50 mL of denatured alcohol, and add 50 mL of water.

3. To prepare 1 L of 0.1 M sodium hydroxide, observe the required safety precautions. Dissolve 4.0 g of NaOH in enough water to make 1 L of solution.

4. To prepare 1 L of 0.1 M sodium chloride, dissolve 5.3 g of NaCl in enough water to make 1 L of solution.

Corrosion is a chemical reaction in which a metal is oxidized. Iron corrodes in the presence of oxygen and water. The metal is converted to brittle metal oxides called rust. Bicycles, car bodies, tools, and appliances become useless and must be replaced when they corrode.

The exact nature of the corrosion process is not well understood, but it is known to be an oxidation-reduction reaction in which iron is initially converted to iron(II) ions.

$$Fe(s) \rightarrow Fe^{2+}(aq) + 2e^-$$

At the same time, hydroxide ions (OH^-) are formed from water and oxygen molecules.

$$\tfrac{1}{2} O_2(g) + H_2O(l) + 2e^- \rightarrow 2OH^-(aq)$$

Adding the two equations gives the following oxidation-reduction equation:

$$Fe(s) + \tfrac{1}{2} O_2(g) + H_2O(l) \rightarrow Fe^{2+}(aq) + 2OH^-(aq)$$

Further reaction with oxygen and water produces rust, so the presence of Fe^{2+} ions and OH^- ions is evidence that corrosion has occurred. The formation of a blue precipitate when $K_3Fe(CN)_6$ is added to a solution indicates the presence of Fe^{2+} ions, as you found in Experiment A4. Phenolphthalein indicator solution turns pink in the presence of OH^- ions. In this experiment, you will use $K_3Fe(CN)_6$ and phenolphthalein to verify the presence of Fe^{2+} and OH^- ions as you investigate some of the factors that affect the process of corrosion.

SAFETY

Always wear safety goggles and a lab apron to protect your eyes and clothing. If you get a chemical in your eyes, immediately flush the chemical out at the eyewash station while calling to your teacher. Know the location of the emergency lab shower and the eyewash station and the procedure for using them.

5. To prepare 1 L of 0.1 M sodium carbonate, dissolve 11 g of Na_2CO_3 in enough water to make 1 L of solution.

6. To prepare 1 L of 0.1 M sodium triphosphate, dissolve 38 g of $Na_3PO_4 \cdot 12H_2O$ in enough water to make 1 L of solution.

7. To prepare 1 L of 0.1 M potassium nitrate, dissolve 10 g of KNO_3 in enough water to make 1 L of solution.

8. To prepare 1 L of 0.1 M sodium oxalate, dissolve 13 g of $Na_2C_2O_4$ in enough water to make 1 L of solution.

 Do not touch any chemicals. If you get a chemical on your skin or clothing, wash the chemical off at the sink while calling to your teacher. Make sure you carefully read the labels and follow the precautions on all containers of chemicals that you use. If there are no precautions stated on the label, ask your teacher what precautions you should follow. Do not taste any chemicals or items used in the laboratory. Never return leftovers to their original containers; take only small amounts to avoid wasting supplies.

 Call your teacher in the event of a spill. Spills should be cleaned up promptly, according to your teacher's directions.

 When you use a candle, confine long hair and loose clothing. If your clothing catches on fire, WALK to the emergency lab shower and use it to put out the fire. Do not heat glassware that is broken, chipped, or cracked. Use tongs or a hot mitt to handle heated glassware and other equipment because hot glassware does not always look hot.

 Never put broken glass into a regular waste container. Broken glass should be disposed of properly.

MATERIALS

9. To prepare 1 L of 0.1 M hydrochloric acid, observe the required safety precautions. Slowly and with stirring, add 8.6 mL of concentrated HCl to enough water to make 1 L of solution.

10. To prepare 1 L of 0.1 M nitric acid, observe the required safety precautions. Slowly and with stirring, add 6.4 mL of concentrated HNO_3 to enough water to make 1 L of solution.

11. To prepare 1 L of 0.1 M sulfuric acid, observe the required safety precautions. Slowly and with stirring, add 5.6 mL of concentrated H_2SO_4 to enough water to make 1 L of solution.

Required Precautions

• Read all safety precautions, and discuss them with your students.

- 0.10 M $K_3Fe(CN)_6$
- agar-agar, powdered
- beaker, 400 mL
- Bunsen burner and related equipment
- copper strip or wire
- glass stirring rod
- iron nails, 14
- iron ring
- magnesium strip
- Petri dishes, 2
- pH paper, wide range
- phenolphthalein indicator
- pliers
- ring stand
- test solutions, 10
    - 0.1 M H_2SO_4
    - 0.1 M HCl
    - 0.1 M HNO_3
    - 0.1 M KNO_3
    - 0.1 M $Na_2C_2O_4$
    - 0.1 M Na_2CO_3
    - 0.1 M Na_3PO_4
    - 0.1 M NaCl
    - 0.1 M NaOH
    - distilled water
- test tubes, 13 mm × 100 mm, 10
- test-tube rack
- tongs
- wire gauze, ceramic center

PROCEDURE

- Safety goggles and a lab apron must be worn at all times.

- In case of an acid or base spill, first dilute with water. Then, mop up the spill with wet cloths or a wet cloth mop while wearing disposable plastic gloves. Designate separate cloths or mops for acid and base spills.

- Wear safety goggles, a face shield, impermeable gloves, and a lab apron when you prepare the NaOH, HCl, HNO₃, and H₂SO₄ solutions. When you prepare the acids, work in a hood known to be in operating condition and have another person present nearby to call for help in case of an emergency. Be sure you are within a 30 s walk of a safety shower and eyewash station known to be in good operating condition.

Procedural Tips

- Review the concepts of oxidation-reduction reactions. Remind students of the Prussian blue test for iron ions, and review how phenolphthalein indicator distinguishes between acids and bases. Show students how to use the pliers to bend the nails.

- Caution students to use beaker tongs or hot mitts to move the beaker containing the suspension of agar-agar.

- Students should not be expected to gain a detailed understanding of the complex process of corrosion, but they will gain an understanding of controlling variables and inferring conclusions from data.

- Students may be interested in hearing about the enormous cost of corrosion in the United States each year—more than 10 billion dollars.

Part 1

1. Label 10 test tubes with the formulas of the test solutions, and put about 5 mL of each solution into the appropriate test tube.

2. Determine the pH of the solutions by dipping a stirring rod into each solution in turn and touching the stirring rod to wide-range pH paper. Be sure to rinse and dry the stirring rod after testing each solution. Record your results in the Data Table.

3. Put a shiny nail in each of the test tubes, and set them aside. Wash your hands.

4. After the nails have been in the test solutions for a day, observe any changes that have taken place. Record the results in the Data Table.

5. Add 1 drop of 0.1 M $K_3Fe(CN)_6$ to each of the test tubes. Record the results in the Data Table.

6. Dispose of the nails as directed by your teacher. Clean and rinse the test tubes, and wash your hands.

Part 2

1. In a 400 mL beaker, heat 200 mL of distilled water to boiling. Add with stirring 2 g of agar-agar. Heat and continue to stir until the agar-agar is thoroughly dispersed. Turn off the heat.

2. With tongs, remove the agar-agar mixture from the ring stand. Add 10 drops of 0.1 M $K_3Fe(CN)_6$ and 5 drops of phenolphthalein indicator. Set the beaker aside to cool.

3. While the agar-agar suspension is cooling, prepare four nails. Use the pliers to bend one nail so that it makes a 90° angle. Wrap a second nail with copper wire and a third with magnesium ribbon.

4. Place the straight nail and the bent nail in one of the petri dishes. Place the nail wrapped with copper and the nail wrapped with magnesium ribbon in the other petri dish. Be sure the nails do not touch each other. Pour the lukewarm agar-agar over the nails until they are covered.

5. Cover the dishes, and set them aside until tomorrow. Wash your hands.

6. The next day, record your observation under *Day 2 Observations*. Remember that all parts of the experiment were allowed to react for the same length of time. Therefore, a greater amount of corrosion indicates an increased rate of reaction.

Cleanup and Disposal

7. Clean all apparatus and your lab station. Return equipment to its proper place. Dispose of chemicals and solutions in the containers designated by your teacher. Do not pour any chemicals down the drain or in the trash unless your teacher directs you to do so. Wash your hands thoroughly after all work is finished and before you leave the lab.

Disposal

The metals may be saved for reuse. If they are too corroded, put them in the trash. Combine all solutions and adjust the pH to be between 5 and 9, then dilute 100-fold and pour down the drain.

Data Table

Solution	pH range	Appearance	$K_3Fe(CN)_6$ test
NaOH	base		
NaCl	no change	corrosion	+
HCl	acid	corrosion	+
Na_2CO_3	base		
H_2SO_4	acid	corrosion	+
$Na_2C_2O_4$	base		
KNO_3	no change	corrosion	+
HNO_3	acid	corrosion	+
Na_3PO_4	base		
Distilled water	no change	corrosion	+

Day 2 observations:

Student answers will vary.

_____

_____

_____

QUESTIONS

1. **Organizing Data** Divide the solutions in Part I into three categories according to their pH.

Acidic	Neutral	Basic
HCl	NaCl	NaOH
H_2SO_4	KNO_3	$Na_2C_2O_4$
HNO_3	H_2O	Na_2CO_3
		Na_3PO_4

2. **Analyzing Results** In which group of solutions was there the most evidence of corrosion?

   the acids

3. **Analyzing Results** In which group of solutions was there the least evidence of corrosion?

   the bases

ChemFile

4. **Relating Ideas** Use the principles of equilibrium to explain your answers to Questions **2** and **3**. (Hint: Refer to the equation in the Introduction. Assuming equilibrium, how would the presence of either H^+ ions or OH^- ions affect the amount of product formed?)

If the corrosion is occurring in a basic environment in which there

are extra OH^- ions, the equilibrium will be shifted to the left.

Fewer $Fe^{2+}(aq)$ and $2OH^-(aq)$ ions will be formed. If the corrosion

is occurring in an acidic environment, the H_3O^+ ions will react

with OH^- ions and shift the equilibrium to the right, the direction

of increased oxidation.

5. **Analyzing Methods** Why was the straight nail included in Part II of the experiment?

To serve as a control for comparison with the bent nail.

6. **Analyzing Results and Relating Ideas** Why do you think the head and the point of the nail corroded more than the middle? (Hint: Think about how nails are made.)

The head and the point have been subjected to more stress in

manufacturing.

7. **Analyzing Results** What regions of the bent nail corroded?

the head, the tip, and the bend

8. **Relating Ideas** Use the Activity Series Table in your textbook to explain why the reactions were different when the nails were wrapped in copper and magnesium.

Magnesium loses electrons more easily than iron, so the

magnesium corrodes instead of the iron. Copper loses electrons

less easily than iron, so the iron corrodes.

9. **Analyzing Ideas**   Explain why corrosion occurs more easily in these regions.

The nail has been subjected to stress in these regions.

10. **Relating Ideas**   Magnesium or zinc blocks are often attached to the hulls of ships. Explain how this helps to prevent corrosion of the ship.

Both zinc and magnesium are more active metals than iron and

will corrode first, or be sacrificed.

GENERAL CONCLUSIONS

1. **Organizing Conclusions**   Summarize the factors that promote corrosion.

acid environment, stress, presence of a less reactive metal

2. **Predicting Outcomes**   The installer of a new battery in your car carelessly splashes some of the battery acid on the inside of your car fender. If the fender is made of iron, should you worry about corrosion? Explain.

yes, because an acid environment promotes corrosion

3. **Inferring Conclusions**   List some actions that might be taken to prevent or reduce corrosion of iron.

• Coat the metal with paint or oil to prevent oxygen and moisture

from reaching it.

• Keep the environment basic.

• Avoid denting and bending the metal.

• Attach the metal to a more active metal.

EXPERIMENT **A23**

Carbon

OBJECTIVES

Recommended time:
60 min if the carbon models are made as a demonstration

- **Construct** models of diamond and graphite.
- **Compare** the observed physical properties of amorphous carbon with its molecular structure.
- **Infer** why activated carbon is a good adsorbing agent.
- **Define** the terms *amorphous carbon* and *allotropic carbon*.

INTRODUCTION

Solution/Material Preparation

1. To prepare 1 L of 1.0 M hydrochloric acid, observe the required safety precautions. Slowly and with stirring, add 86 mL of concentrated HCl to enough distilled water to make 1 L of solution.

2. To prepare 1 L of 1.0 M sodium hydroxide solution, observe the required safety precautions. Slowly and with stirring, add 40 g of NaOH to enough distilled water to make 1 L of solution.

Carbon exhibits allotropy, the existence of an element in two or more forms in the same physical state. Diamond and graphite are carbon's two crystalline allotropic forms. They result from the different ways in which carbon atoms link to one another in the two forms.

Microscopic graphite structures are contained in the black residues known as amorphous carbons that are obtained by heating certain carbon substances. Examples of amorphous carbon include charcoal, coke, bone black, and lampblack. The amorphous forms are produced by a variety of procedures, including destructive distillation and incomplete combustion. If a carbon-rich compound such as gasoline, C_8H_{18}, is burned completely, the only products will be carbon dioxide and water vapor.

$$2C_8H_{18}(l) + 25O_2(g) \rightarrow 16CO_2(g) + 18H_2O(g)$$

Unfortunately, most internal combustion engine that run on gasoline are far from perfect. Incomplete combustion often results in black smoke (partially burned hydrocarbons) and carbon monoxide in the exhaust gases.

Carbon is an important reducing agent. The uses of its different varieties are related to particular properties, including combustibility and adsorption.

SAFETY

3. To prepare 1 L of dark-brown sugar solution, add 5 g of dark-brown sugar to enough water to make 1 L of solution.

Always wear safety goggles and a lab apron to protect your eyes and clothing. If you get a chemical in your eyes, immediately flush the chemical out at the eyewash station while calling to your teacher. Know the location of the emergency lab shower and the eyewash station and the procedure for using them.

Do not touch any chemicals. If you get a chemical on your skin or clothing, wash the chemical off at the sink while calling to your teacher. Make sure you carefully read the labels and follow the precautions on all containers of chemicals that you use. If there are no precautions stated on the label,

HRW material copyrighted under notice appearing earlier in this work.

4. To prepare 1 L of limewater, add 5 g of Ca(OH)$_2$ to 1 L of distilled water. Shake well and allow the undissolved solids to settle. Decant the saturated solution into another bottle.

Required Precautions

• Read all safety precautions, and discuss them with your students.

• Safety goggles and a lab apron must be worn at all times.

ask your teacher what precautions you should follow. Do not taste any chemicals or items used in the laboratory. Never return leftovers to their original containers; take only small amounts to avoid wasting supplies.

 Call your teacher in the event of a spill. Spills should be cleaned up promptly, according to your teacher's directions.

 When using a Bunsen burner, confine long hair and loose clothing. If your clothing catches on fire, WALK to the emergency lab shower, and use it to put out the fire. Do not heat glassware that is broken, chipped, or cracked. Use tongs or a hot mitt to handle heated glassware and other equipment because hot glassware does not always look hot.

 Never put broken glass into a regular waste container. Broken glass should be disposed of properly.

MATERIALS

• In case of a spill, use a dampened cloth or paper towels to mop up the spill. Then rinse the cloth in running water at the sink, wring it out until it is only damp, and put it in the trash. In the event of an acid or base spill, dilute with water first, and then proceed as described.

- 1.0 M HCl
- 1.0 M NaOH
- 50 mL graduated cylinder
- 250 mL beaker
- 125 mL Erlenmeyer flask
- activated charcoal, (for decolorizing)
- balance
- Bunsen burner and related equipment
- buret clamp
- charcoal wood splinters
- crucible and cover
- dark-brown sugar solution
- filter paper

- forceps
- funnel
- funnel rack
- iron ring
- limewater
- molecular model kits, 2
- pipe-stem triangle
- test tube, 13 mm × 100 mm
- test tube, 25 mm × 100 mm, 6
- ring stand
- rubber stopper, solid No. 4
- rubber stoppers, solid No. 2, 5
- sugar, white granular
- test-tube holder
- wooden splints

PROCEDURE

• Wear safety goggles, a face shield, impermeable gloves, and a lab apron when you prepare the 1.0 M HCl and the 1.0 M NaOH. When making the HCl solution, work in a hood known to be in operating condition and have another person stand by to call for help in case of an emergency.

1. Use a molecular model kit to start on the construction of the graphite model by constructing two hexagons using only the short connectors. Recall that the distances between the centers of adjacent carbon atoms in a layer of graphite are identical (142 pm). With Figure A as a guide, connect the two hexagons using the longer stick connectors to represent the distance between centers in adjacent layers (335 pm). Use the remaining carbon spheres to build up each layer. Two carbons in the first hexagons you constructed will be common to the newly constructed hexagons. Use long connectors as needed between the rest of the constructed layers.

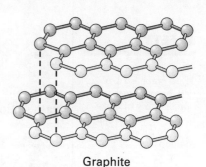

Graphite

FIGURE A

Be sure you are within a 30 s walk of a safety shower and eyewash station known to be in good operating condition.

Procedural Tips

• To save time, the models could be assembled in a demonstration.

• Students should understand that incomplete combustion of carbon-containing substances results when not enough oxygen is available to form CO_2 with all of the carbon in the substance. Some of the carbon combines to form CO; some doesn't combine at all, and therefore elemental carbon is produced.

Disposal

• Combine all the liquids. Adjust the pH to between 5 and 9 and pour down the drain. Collect all solids; ensure that there are no flames or burning embers, and put into the trash.

2. Use 20 spheres representing carbon atoms from the molecular model kit to construct a model of diamond. Connect the spheres by means of the short stick connectors in the arrangement of atoms shown in Figure B. How many carbon atoms is each carbon bonded to in the hexagon layer of graphite? What is the nature of the bonding between layers in graphite? How many carbon atoms is each carbon bonded to in diamond? Which structure is more rigid?

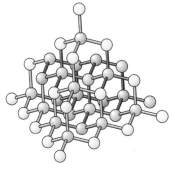

Diamond

FIGURE B

Observations:

Each carbon is bonded to three carbon atoms in the hexagon

layer. The layers are held together by weak dispersion forces.

Carbon is bonded to four other atoms in diamond. Diamond

is the more rigid structure.

3. Ignite a wood splint in the burner flame. Note how it burns.

4. Break several splints and place the pieces in the bottom of a test tube. Clamp the test tube to the ring stand so that its mouth is slightly downward, as shown in Figure C. Place a piece of paper on the table under the mouth of the test tube and heat the contents strongly until no more volatile matter is released.

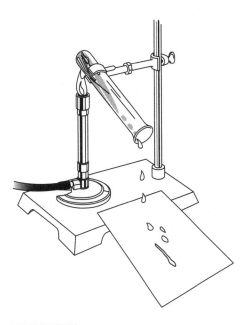

FIGURE C

HRW material copyrighted under notice appearing earlier in this work.

5. Remove the residue from the test tube carefully. Describe all the products of the reaction you observed.

Observations:

Liquid droplets from the tube fall on the paper. The residue in the tube is charcoal black.

6. Using forceps, ignite one piece of the solid residue in the burner and observe how it burns.

Observations:

student observations

7. Pour one inch of limewater into a small test tube. Using forceps, ignite another piece of solid residue, and insert the ignited end into the test tube, holding it above the limewater. Remove the ignited piece. Immediately stopper and shake the test tube. If nothing happens to the limewater, repeat with another piece of ignited residue. What happens to the limewater? What does this indicate about the nature of the residue?

Observations:

The limewater turns cloudy, indicating the presence of CO_2 gas.

8. Conduct an experiment to show that sugar may be decomposed into carbon and water vapor. Place 2 g of sugar into a porcelain crucible and cover it. Set the crucible and contents in a pipe-stem triangle and heat.

Observations:

student observations

9. Using a test-tube holder, hold a test tube horizontally with its closed end in the yellow burner flame. What form of carbon is produced?

Observations:

Carbon black is deposited on the test tube.

10. Change to the blue flame and hold the test tube so that the black deposit is in the hottest part of the flame.

Observations:

The deposit is removed by the oxidizing blue flame.

CAUTION Be careful when handling the acids and bases in Procedure step 11. Avoid contact with skin and eyes. If any of these chemicals should spill on you, immediately flush the area with water and then notify your teacher.

11. Conduct experiments to determine the activity and solubility of wood charcoal splinters in dilute hydrochloric acid, water, and sodium hydroxide solution. Using three labeled test tubes, place charcoal splints into 5 mL of each of the solutions. Stopper the test tubes and shake the contents. Describe your results. Dispose of the liquids and any pieces of charcoal as directed by your teacher.

Observations:

There was no evidence of reactions taking place in any of the solutions.

12. Pour 50 mL of dark-brown sugar solution into a 125 mL Erlenmeyer flask. Add 1 g of powdered activated charcoal and stopper the flask. Shake the mixture vigorously for 1 min.

13. Set up a funnel with filter paper, and filter the mixture. Pour the filtrate through the funnel a second time, and even a third time, if necessary, to remove the color from the sugar solution. Finally, compare the color of the filtrate with that of the original brown-sugar solution.

Observations:

The filtrate is colorless.

Cleanup and Disposal

14. Clean all apparatus and your lab station. Return equipment to its proper place. Dispose of chemicals and solutions in the containers designated by your teacher. Do not pour any chemicals down the drain or in the trash unless your teacher directs you to do so. Wash your hands thoroughly before you leave the lab and after all work is finished.

QUESTIONS

1. **Inferring Conclusions**   Of what use is activated charcoal in water purification processes?

   Charcoal removes objectionable color and odors from water.

2. **Analyzing Information**   What is adsorption? Why are certain forms of charcoal good adsorbing agents?

   Adsorption is the concentration of a gas, liquid, or solid on the

   surface of a liquid or solid with which it is in contact. Activated

   charcoal is prepared in such a way that it has very large internal

   surface areas. This makes it an efficient adsorbing agent.

GENERAL CONCLUSIONS

1. **Analyzing Results**   In this experiment you used several different methods to prepare samples of carbon. What properties do all of the samples have in common? How does the model you built in step **1** explain these properties?

   All samples are unreactive, black, and somewhat soft.

   Graphite is a network structure and is arranged in layers.

2. **Analyzing Ideas**   Automobile engineers use the ratio of carbon monoxide to carbon dioxide in exhaust gases as a measure of the efficiency of an engine. Which is better, a high ratio or a low ratio? Explain your answer. (Hint: Refer to the equation for the complete combustion of gasoline in the Introduction.)

   A low ratio is better. The less carbon monoxide and the more

   carbon dioxide in the exhaust, the more complete the combustion.

EXPERIMENT **A24**

Oil-Degrading Microbes

OBJECTIVES

- **Compare** the physical characteristics of oil before and after the action of oil-degrading microbes.
- **Identify** which microorganisms are useful in cleaning up an oil spill.
- **Graph** the change in turbidity for each system.
- **Apply** aseptic techniques in a laboratory procedure.

INTRODUCTION

Recommended time:
One 50-minute period to set up experiment and 10–15 minutes over the next four to seven days to collect data. *Note: Cultures grown at 30°C will achieve faster results than those grown at a cooler temperature. Length of the experiment may need to be adjusted for this variable.*

Through the news media, the public has been made increasingly aware of the hazards to the environment caused by oil spills. A single gallon of oil can spread thinly enough to cover four acres of water. Some of the oil evaporates; some is broken down by radiant energy; and some emulsifies, or breaks down into small pieces, to form a heavy material that eventually sinks to the bottom of the ocean. This heavy material endangers birds, marine mammals, and other forms of sea life.

Oil spills can be cleaned up by mechanical techniques, such as the use of skimmers and barriers, as well as by using chemical dispersants and solvents and by burning the oil. In addition, some microorganisms can break down the various hydrocarbons in an oil spill. This may be the best, most environmentally safe prospect for cleaning up oil spills. These "oil-hungry" microbes convert oil into food for themselves, rendering the oil nontoxic and allowing it to be assimilated safely into the aquatic food web. The use of living organisms to repair environmental damage is known as **bioremediation.**

You are an environmental chemist who works for a large petroleum company. Your job is to research new methods of cleaning up oil spills. You have read that certain kinds of microbes digest oil. You want to find out if these microbes can be used to digest oil from an oil spill. You will set up a controlled experiment to find out which of two different microbes digests oil more efficiently.

SAFETY

For this experiment wear protective gloves. Always wear safety goggles and a lab apron to protect your eyes and clothing. If you get a chemical in your eyes, immediately flush the chemical out at the eyewash station while calling to your teacher. Know the location of the emergency lab shower and the eyewash station and the procedure for using them.

Do not touch any chemicals used in the laboratory. If you get a chemical on your skin or clothing, wash the chemical off at the sink while calling to your teacher. Make sure you carefully read the labels and follow the precautions on all containers of chemicals that you use. If there are no

Solution/Material Preparation

1. Materials for this activity can be purchased from WARD'S Natural Science Establishment, Inc.

MATERIALS

5100 W. Henrietta Road
P.O. Box 92912
Rochester, NY
14692-9012
Penicillium (order number 85 T 7110)
Pseudomonas
(85 T 1704)
Nutrient broth
(88 T 0503)
Biohazard bag, 100 ct
(18 T 6905)
or Kit: Oil-Degrading
Microbes Oil Spill
(85 T 3503)

PROCEDURE

2. Rehydrate the growth culture four to five days prior to use in the investigations. You may also have your students participate in the procedures of rehydration and growing of cultures. Rehydration instructions come with the special oil-degrading, freeze-dried *Pseudomonas* and *Penicillium* cultures.

Required Precautions

• Read all safety precaution and discuss them with your students.

• Although these oil-degrading microbes are naturally occurring and nonpathogenic, tell students they must wear safety goggles, a laboratory apron, and gloves throughout the procedure to minimize the risk of contamination.

precautions stated on the label, ask your teacher what precautions to follow. Never return leftovers to their original containers; take only small amounts to avoid wasting supplies.

 Call your teacher in the event of a spill. Spills should be cleaned up promptly, according to your teacher's directions.

• protective gloves
• plastic jars with lids, 3
• wax pencil
• 90 mL distilled water
• 1.5 g nutrient fertilizer
• density indicator strips, 3
• disposable pipets, 3
• *Pseudomonas* culture
• *Penicillium* culture
• refined oil in a dropper bottle
• disinfectant solution in a squeeze bottle
• paper towels
• scoop
• biohazard waste-disposal container

Part 1—Inoculating Oil with Microorganisms

1. Use disinfectant solution to sterilize the top of your work bench. To do this, spread disinfectant solution over the entire work area. Wipe the area clean with paper towels. Throw paper towels into trash barrels indicated by your teacher. Why is it important to sterilize your work area before you begin this investigation?

Sterilization is important to prevent contamination of the test

cultures from microorganisms that might be present on the work

bench.

_____

_____

2. Use a wax pencil to label the first plastic jar "Control," the second plastic jar *"Pseudomonas,"* and the third plastic jar *"Penicillium."* Also write the date and your name and class period on each jar.

3. Pour about 30 mL of distilled water into each jar so that each jar is about half full.

4. Add about 20 drops of refined oil to form a thin layer in each jar.

5. Using a scoop, add a pinch of fertilizer to the water mixture. This fertilizer is the kind used in real oil spills; it will coagulate the oil and provide nutrients for microbial growth.

- Review the safety precautions for handling microorganisms in the lab. Remind students to use aseptic technique at all times during this lab.

- Make sure students wash their hands with antibacterial soap before and after the lab and sterilize their work areas before and after the investigation.

Procedural Tips

- Divide the class into teams of two to four students.

- Set up culture-growing stations, with a plastic pipet at each culture that students can use to inoculate their test tubes.

- Although specific observations may vary from student to student, they should agree that the oil starts showing signs of degradation after one to two days of microbial action, based on its color change from deep brown to yellowish brown and the changes in its physical characteristics. The turbidity of the solutions should increase, indicating microbial growth. In this case, the fertilizer adds an additional carbon source (nutrients) for the microbes.

- The refined oil used in this investigation is light brown in color and forms a smooth and continuous layer on the water's surface.

Disposal

- When students have completed this lab, autoclavable materials should be sterilized either in an autoclave or pressure cooker. All materials that came in contact with bacteria should be autoclaved at 121°C and 15 psi for 20 minutes before disposal.

6. Inoculate the appropriate jars with 5 mL of the *Pseudomonas* culture and 5 mL of the *Penicillium* culture. The control jar receives no microorganisms. **CAUTION: Always practice aseptic technique when using bacteria in the lab.**

7. Secure the top on each jar, and invert the jar several times to mix the contents. Similar wave action occurs in the ocean and increases the amount of dissolved oxygen in the water. Why is it desirable to increase the amount of dissolved oxygen in the water?

The microorganisms will grow more readily in water with a high

oxygen concentration.

8. Number the bars on three density indicator strips from 1 to 5, with 5 for the darkest. Place a density strip on each jar so that the tops of the bars are below the water level. The density strips will allow you to quantify the amount of microbial growth in each jar. As the microbes grow, the water will become cloudy, or turbid. As the microbe population increases, you will be able to see fewer bars through the water. When viewing the jars, place a piece of white paper behind each jar to provide a background. Look through the water in the jar to determine the number of density bars that have disappeared.

9. Record your observations for today, day 0, in the Data Table.

10. Incubate the jars with caps half loosened at 30°C. If an incubator is not available, place the jars in a warm spot in the classroom as directed by your teacher. Why is it necessary to incubate the jars?

Incubation is necessary to allow the organisms time to multiply.

A well-developed culture will degrade the oil more effectively.

Why is it necessary to leave the jar caps half loosened during incubation?

The jar caps must remain loosened to allow movement of oxygen

and carbon dioxide into and out of the jar.

Cleanup and Disposal

11. Dispose of your materials according to the directions from your teacher. Sterilize your work area as described in step 1. Wash your hands with antibacterial soap before leaving the lab.

Part 2—Observing the Effects of Microbes on Oil

12. Observe the jars every 24 hours for four days. Look for any signs of oil degradation, such as change in color, formation of tiny oil droplets, break-up of the oil layer into smaller fragments, or changes in texture. Microbial growth will be observed mainly at the interface (boundary) between the oil and water. Record your observations of the general appearance of the degrading oil, including its color and the turbidity of the water in the Data Table. Each day after you make your observations, invert each jar once or twice to increase the dissolved oxygen content in the water and to further mix the oil with the jar's other ingredients.

Which microorganisms do you think will degrade the most oil?

Answers will vary.

Data Table

Day	Organism	General appearance of oil	Color of oil	Turbidity of water from density strip (number of bars that *disappear*)
0	Pseudomonas	large blob in the middle of the jar	light brown	1
	Penicillium	large blob in the middle of the jar	light brown	1
	Control	large blob in the middle of the jar	light brown	1
1	Pseudomonas	large blob in the middle of the jar	not as dark as day 0	2
	Penicillium	large blob in the middle of the jar	not as dark as day 0	2
	Control	large blob in the middle of the jar	light brown	1
2	Pseudomonas	smaller blobs of oil	yellowish brown	2
	Penicillium	smaller blobs of oil	yellowish brown	2
	Control	large blob in the middle of the jar	light brown	1
3	Pseudomonas	oil blobs do not appear as thick	ivory	3
	Penicillium	oil blobs do not appear as thick	ivory	3
	Control	large blob in the middle of the jar	deep brown	1
4	Pseudomonas	thin drops	ivory	3
	Penicillium	thin drops	ivory	3
	Control	large blob in the middle of the jar	light brown	1

QUESTIONS

• Solid trash that has been contaminated by biological waste must be collected in a separate and specially marked biohazard bag. Package all sharp instruments in separate metal containers for disposal. Pipets should not protrude through the disposal bag. After autoclaving, disposal bags should be placed in a sealed container (plastic bucket with lid).

1. **Organizing Data** Prepare a graph showing the change in turbidity that occurs in each jar. The turbidity (number of bars that disappear) should be shown on the *y*-axis, with time plotted on the *x*-axis. Use different-colored pencils to draw the curve for growth of both organisms and the control on the same graph. Make a legend that indicates the colors that represent the organisms and the control.

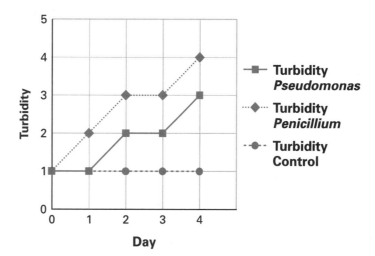

2. **Analyzing Results** Describe any changes in the physical characteristics and appearance of the oil on day 1 and beyond. Discuss the possible causes of such changes.

The oil should start to show signs of degradation after one or two days of microbial action, changing color from dark brown to yellowish-brown and eventually to ivory. Another change in its physical characteristics is the breakdown of the continuous layer of oil on the water's surface into minute oil droplets. The control jar should show no change.

3. **Analyzing Results** What does an increase in turbidity indicate?

An increase in turbidity indicates an increase in microbial population as well as degradation of the oil.

4. **Analyzing Methods** What is the purpose of the control?

The control shows that there was a change in the oil—in both color and texture—throughout the experiment. The color of the oil became lighter, and the oil broke down into smaller pieces.

5. **Analyzing Methods** What is the purpose of starting with distilled water rather than tap water or pond water?

Tap water may have chlorine in it that could kill the bacteria.

Pond water will contain nutrients and other bacteria, adding

more variables to this investigation.

6. **Analyzing Methods** How does the procedure you used in this lab differ from an actual oil spill?

Answers will vary but may include the following ideas:

Salt water was not used; no existing bacteria or other organisms

were in the water; no wave action constantly stirred up the water

and oxygenated it; and no sand, rocks, or other shoreline material

was present.

GENERAL CONCLUSIONS

1. **Predicting Outcomes** What would happen to the growth of the microbes if no fertilizer were added?

The microbes would still grow and degrade the bacteria, but

probably at a slower rate. The fertilizer helped coagulate the oil

and added nutrients that would allow the microbes to grow and

reproduce faster.

2. **Inferring Conclusions** Which microbe degraded the oil better?

In general, both organisms degraded the oil at the same rate.

Some students will note that the *Penicillium* culture created more

turbid water, but this is due to the formation of hyphae, not

increased rate of degradation.

3. **Applying Ideas** What advantages, if any, can you think of for using a mixture of microorganisms rather than just one kind of microorganism to degrade the oil?

By using a mixture of microorganisms, a broad spectrum of

hydrocarbon degradability can be attained, something that is

not possible by using only one type of microorganism.

EXPERIMENT **A25**

Polymers

OBJECTIVES

Recommended time:
60 min

- **Infer** chemical structures from differences in chemical properties.
- **Deduce** the solubility of plastic foam.
- **Describe** the properties of three polymers.

INTRODUCTION

Solution/Material Preparation

1. Break a foam cup into pieces or use packing "peanuts." One or two cups should be enough for a class.

2. To prepare 1 L of 4% polyvinyl alcohol, slowly add 40 g of solid polyvinyl alcohol to about 960 mL of water while stirring with a stirrer. Then heat the mixture to about 70°C and continue stirring at this temperature for about an hour or until the solution is clear. Some chemical suppliers provide prepared solutions.

3. Sodium polyacrylate can easily be dispensed from small, well-labeled salt shakers. Avoid breathing the dust when filling the shakers by wearing a dust mask. Also wear the dust mask when preparing the sodium borate solution. To dispose of the dust mask, wrap it in a dampened newspaper and put it into the trash.

Polymers are giant molecules consisting of repeating groups of atoms called monomers, which form chains that are thousands of atoms long. Because these long molecular chains are linked by intermolecular forces, they can be molded into useful objects. If short bridges of atoms form between long polymeric chains the polymer is then said to be cross-linked. Cross-linking gives the polymer new properties. In this experiment, you will investigate the properties of polystyrene, polyvinyl alcohol, and sodium polyacrylate.

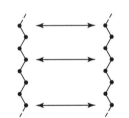

Intermolecular Forces of Attraction

Cross-linking

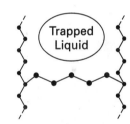

Gel Formation

Polystyrene is not cross-linked, but its intermolecular forces of attraction make it useful for constructing products such as radio cases, toys, and lamps. When polystyrene is expanded to produce the material called plastic foam, it has a very low density and is used to make egg cartons, insulation, and fast-food containers. The intermolecular forces of attraction in polystyrene are destroyed by the action of the acetone and the polymer loses its shape and becomes fluid.

Polystyrene Chain

4. To prepare 1 L of 4% sodium borate, dissolve 40 g of solid borax, $Na_2B_4O_7 \cdot 10H_2O$, in enough water to make 1 L of solution.

Required Precautions

• Read all safety precautions, and discuss them with your students.

• Safety goggles and a lab apron must be worn at all times.

• In case of a spill, use a dampened cloth or paper towels. Then rinse the cloth in running water at the sink, wring it out until it is only damp, and put it in the trash. In the event of an acid or base spill, dilute with water first, and then proceed as described.

Polyvinyl alcohol can be weakly cross-linked with the hydrated borate ion. This polymer forms a non-Newtonian gel, which has properties similar to the Slime toy manufactured by Mattel Toy Corporation. When kept in motion, it forms a semi-rigid mass; when held steady, it flows.

Polyvinyl Alcohol Cross-linked with Borate Ion

Sodium polyacrylate is a strongly cross-linked polymer that has superabsorbent properties. It can form a gel by absorbing as much as 800 times its mass of water. Currently it is used to coat seeds before planting and to remove water from diesel and aviation fuels. Some brands of disposable diapers contain this superabsorbent polymer.

SAFETY

• Acetone is flammable. Make sure there are no flames or other sources of ignition present before using this chemical. Limit the amount of acetone in the laboratory to no more than 200 mL in two or more bottles, each containing no more than 100 mL.

• Contrary to common belief, borax is toxic. Do not allow students to handle borate polymer with bare hands. Do not allow students to take even a small sample of the borate polymer out of the lab.

 For this experiment, wear safety goggles, protective gloves, and a lab apron to protect your eyes, hands, and clothing. If you get a chemical in your eyes, immediately flush the chemical out at the eyewash station while calling to your teacher. Know the location of the emergency lab shower and the eyewash station and the procedure for using them.

 Do not touch any chemicals used in the laboratory. If you get a chemical on your skin or clothing, wash the chemical off at the sink while calling to your teacher. Make sure you carefully read the labels and follow the precautions on all containers of chemicals that you use. If there are no precautions stated on the label, ask your teacher what precautions you should follow. Do not taste any chemicals or items used in the laboratory. Never return leftovers to their original containers; take only small amounts to avoid wasting supplies.

 Call your teacher in the event of a spill. Spills should be cleaned up promptly, according to your teacher's directions.

 Never put broken glass in a regular waste container. Broken glass should be disposed of properly.

MATERIALS

Procedural Tips

• Discuss the structures of the polymers given in the Introduction.

• Remind students that they are seeing three-dimensional structures presented in two dimensions.

PROCEDURE

• Encourage them to think about how the polymers might occupy space given the tetrahedral orientation of bonds around each carbon.

Disposal

• Collect the acetone (if any), dilute it 20-fold with water, and pour it down the drain. Collect the foam-acetone pieces. In a hood known to be operating properly, let any residual acetone evaporate, and then put the remainder into the trash. All else can go directly into the trash.

• 4% polyvinyl alcohol
• 4% sodium borate solution
• 10 mL graduated cylinder
• 50 mL beaker
• 5 mL acetone
• foam cup
• glass stirring rod
• medicine dropper
• NaCl
• paper towel
• petri dish
• sodium polyacrylate
• spatula
• squares of paper towel
• watch glass

1. Measure 5 mL of acetone and pour it into a petri dish. Place two or three pieces of expanded polystyrene (from a foam cup) into the acetone as shown in Figure A. After a few minutes, examine the polystyrene.

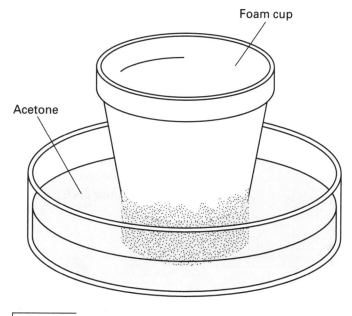

Foam cup

Acetone

FIGURE A

Observations:

The foam pieces slowly lose their shape and settle into blobs.

2. Use a spatula to lift out the polystyrene pieces and place them onto a paper towel. Using a medicine dropper, place a few drops of acetone from the petri dish onto a watch glass and allow it to evaporate. Dispose of the remaining acetone in the metal safety can provided by your teacher.

Observations:

When the solvent evaporated, very little residue remained.

3. Pour 25 mL of 4% polyvinyl alcohol into a 50 mL beaker. With stirring, slowly add 3 mL of 4% sodium borate solution. Be sure you have your gloves on. Pour the gel onto the bench top. Pick it up and knead it into a ball. Pull it slowly, as shown in Figure B. Pull it quickly. Hold part of it in your hand over a beaker and let the rest stretch downward.

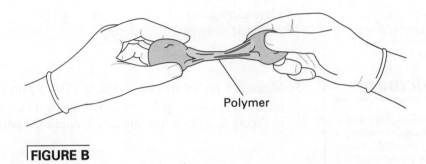

Polymer

FIGURE B

Observations:

When held steady, it flows, and when pulled sharply, it breaks.

CAUTION Although sodium polyacrylate is nontoxic, it readily absorbs water. For this reason, its dust should not be inhaled.

4. Place two 10 cm squares of paper towel on the bench, about 5 cm apart. Lightly sprinkle one of the paper towels with sodium polyacrylate polymer, as shown in Figure C. Cover each paper towel with a second paper towel.

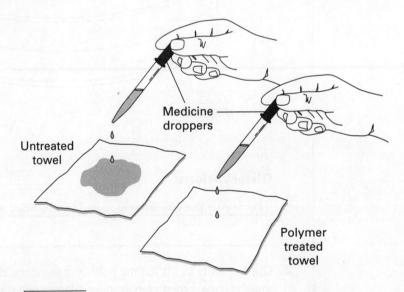

Medicine droppers

Untreated towel

Polymer treated towel

FIGURE C

5. Slowly empty a water-filled medicine dropper onto the center of the first pair of paper towels, and then empty another dropper of water onto the center of the other pair. Repeat this procedure until one pair of the paper towels appears saturated with water. Remove the top paper towels and examine each bottom towel.

Observations:

The polymer on the paper towel absorbed the water by forming

a gel.

6. In the center of a clean Petri dish lid, lightly sprinkle a few grains of sodium polyacrylate polymer. Slowly add a few drops of water until a gel forms. By repeating this process, form a tower of gel 2 cm high. Then sprinkle a few grains of salt onto the top, as shown in Figure D.

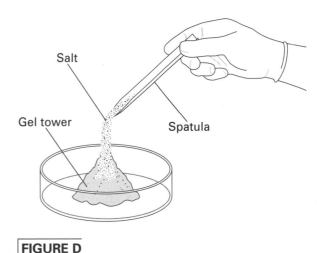

Salt

Gel tower

Spatula

FIGURE D

Observations:

When salt is added to the gel, it becomes very fluid, and the

tower of gel is destroyed.

Cleanup and Disposal

7. Clean all apparatus and your lab station. Return equipment to its proper place. Dispose of chemicals and solutions in the containers designated by your teacher. Do not pour any chemicals down the drain or in the trash unless your teacher directs you to do so. Wash your hands thoroughly after all work is finished and before you leave the lab.

QUESTIONS

1. Analyzing Results What evidence suggests that the foam pieces did not significantly dissolve in the acetone?

When some of the acetone was removed from the Petri dish

and evaporated, very little residue was obtained.

2. **Analyzing Results** The slime that you prepared in step **3** stretches when pulled slowly, but breaks when pulled quickly. How can you explain this behavior based on its structure?

Slow pulling permits the borate ion time to form a cross-link at a

new location.

3. **Analyzing Results** Refer to the structures of the polymers, and explain why sodium polyacrylate is able to absorb large quantities of water but polystyrene does not have this property.

Sodium polyacrylate is strongly cross-linked with the borate ion.

There are internal spaces in the structure where water molecules

can fit. Polystyrene is not cross-linked, so it doesn't have spaces

for absorbed water.

GENERAL CONCLUSIONS

1. **Relating Ideas** What is the advantage of coating a seed with sodium polyacrylate polymer before planting?

This polymer facilitates seed germination by keeping the seed

moist.

2. **Relating Ideas** The polymers you investigated in this lab are only a small fraction of the polymers you encounter everyday. Natural polymers are found in your body, in the foods you eat, and in plant structures. Name some polymers that are familiar to you.

hair and skin, starches, and cellulose and lignin of plants

Name _____

Date _____ Class _____

EXPERIMENT **A26**

Radioactivity

OBJECTIVES

Recommended time:
35 minutes

- **Use** a Geiger counter to determine the level of background radiation.
- **Compare** counts per minute for a beta source passing through air, index cards, and aluminum.
- **Graph** data and determine the relationship between counts per minute and thickness of material.
- **Graph** data and determine the effect of distance on counts per minute in air.

INTRODUCTION

Solution/Material Preparation

1. Thallium 204 is an inexpensive beta source with a useful life of about 15 years. Read the special precautions about radioactive sources.

2. A decade scaler is an ideal radioactive counter. Thorton has an excellent scaler that is one of the least-expensive models available.

The types of radiation emitted by radioactive materials include alpha particles, beta particles, and gamma rays. Alpha particles are helium nuclei; beta particles are high-speed electrons; and gamma rays are extremely high-frequency photons. Alpha particles will discharge an electroscope, but their penetrating power is not great enough to affect a Geiger counter tube. A few centimeters of air will stop an alpha particle. Beta particles can be detected by a Geiger counter tube and can penetrate several centimeters of air, but several layers of paper or aluminum foil can stop them. Gamma rays can penetrate through several centimeters of concrete.

The environment contains a small amount of natural radiation, which can be detected by a Geiger counter. This is called background radiation and is primarily due to cosmic rays from stars. A small amount of background radiation may also come from the walls of buildings made of stone, clay, and some kinds of bricks, and from other sources such as dust particles.

SAFETY

Required Precautions

- Read all safety cautions, and discuss them with your students.

- Safety goggles and a lab apron must be worn at all times.

  For this experiment wear safety goggles, gloves and a lab apron to protect your eyes, hands, and clothing. If you get a chemical in your eyes, immediately flush the chemical out at the eyewash station while calling to your teacher. Know the location of the emergency lab shower and the eyewash station and the procedure for using them.

 Do not touch any chemicals. If you get a chemical on your skin or clothing, wash the chemical off at the sink while calling to your teacher. Make sure you carefully read the labels and follow the precautions on all containers of chemicals that you use. If there are no precautions stated on the label, ask your teacher what precautions you should follow. Do not taste any chemicals or items used in the laboratory. Never return leftovers to their original containers; take only small amounts to avoid wasting supplies.

MATERIALS

- aluminum foil
- index cards, several

- radioactivity demonstrator (scaler)
- thallium 204 (beta source)

PROCEDURE

• Before your students perform this experiment, you should read the manufacturer's precautions for the use of the counter equipment. Be sure you and your students understand and follow all the precautions. Similarly, be sure you and your students will follow the supplier's precautions for the use and handling of the beta source. When not in use, the beta source should be kept secure in a locked location not accessible to students.

Procedural Tips

• Be sure that students understand how to use the counter according to the directions accompanying your counters. Show students how to set up the counter and Beta source for measurement. Require that they measure the distances to ± 1 cm.

• Ask students why it is necessary to determine a background count and subtract it from the values they obtain when measuring the output of the Beta emitter.

• Discuss the graphing of data and what the slope of a straight line means when assessing the dependence of one variable on another.

Disposal

Students must return the beta emitter to you for safe storage as described under Required Precautions.

Part 1: Background Count

CAUTION The wiring between the Geiger counter probe and the counter carries more than 1000 volts. Do not touch it when the equipment is operating or for at least 5 min after it has been unplugged from the electric socket.

1. Carefully read the directions on the operation of your counter. Set the counter to zero. Do not have any radioactive sources within 1 m of the Geiger counter tube. Turn on the counter for 1 min and count the frequency of clicks. Record the number of clicks per minute in Data Table 1. Repeat two more trials. The average of these trials will be your background count in counts per minute (cpm).

Part 2: Range of Beta Particles in Various Media

2. Air: Place the beta source approximately 5 cm from the Geiger tube. The proper setup is shown in Figure A.
CAUTION You should never directly handle a radioactive source. Be certain you are wearing gloves when handling the beta source (thallium 204).
   Determine the count for 1 min, and record your count/per minute and distance between the source and Geiger tube in Data Table 2.

3. Move the source another 5 cm, and determine and record the count for 1 min. Continue this procedure until you reach the background count or have completed 10 trials.

4. Paper: Place the beta source approximately 5 cm from the Geiger tube. Place a single index card on top of the beta source and determine the count for 1 min. Record the number of counts per minute in Data Table 3.

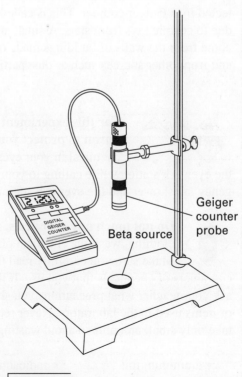

Geiger counter probe

Beta source

DIGITAL GEIGER COUNTER

FIGURE A

ChemFile

5. Keep the distance from the source to the Geiger tube constant, and repeat step **4** with an additional index card. Continue adding index cards and measuring and recording counts until the background count is reached or 10 trials have been completed.

6. Aluminum: Repeat steps **4** and **5,** but use sheets of aluminum foil in place of index cards. Record your data in Data Table 3.

7. Return the beta source to your teacher.

Data Table 1

Background Count

Trial	Count/min
1	
2	
3	
Average	

Data Table 2

Air

Distance (cm)	Count/ min	Distance (cm)	Count/ min	Distance (cm)	Count/ min

Data Table 3

Index Cards		Aluminum	
No. Sheets	Count/min	No. Sheets	Count/min
0		0	
1		1	
2		2	
3		3	
4		4	
5		5	
6		6	
7		7	
8		8	
9		9	
10		10	

1. **Organizing Data** On a separate piece of graph paper, plot a graph of counts per minute versus distance between the beta source and the Geiger tube. Subtract the background count from each of the readings. Place counts per minute on the vertical axis and the distance on the horizontal axis. Describe the relationship between distance and counts per minute.

The graph shows an inverse relationship.

2. **Organizing Data** On a separate piece of graph paper, plot a graph of the number of index cards versus counts per minute. Place the counts per minute on the vertical axis. Describe the relationship between the number of index cards and counts per minute.

The graph shows an inverse relationship.

3. **Organizing Data** On a separate piece of graph paper, plot a graph of the number of sheets of aluminum foil versus counts per minute. Place the counts per minute on the vertical axis. Describe the relationship between the number of sheets of aluminum and counts per minute.

The graph shows an inverse relationship.

**GENERAL
CONCLUSIONS**

1. **Analyzing Results** According to the three graphs, which substance is the most efficient absorber of beta particles? Explain how you made your decision.

Aluminum; the slope of the line is the least when aluminum foil is

used.

2. **Relating Ideas** What is the implication of discovering a larger-than-normal amount of helium in a natural-gas well?

A higher concentration of helium in natural gas could mean that

there is a higher-than-normal concentration of radioactive

material in that location.

Name _____

Date _____ Class _____

EXPERIMENT **A27**

Detecting Radioactivity

OBJECTIVES

Recommended time:
2 lab periods (with exposure time of 3 weeks—can be shortened with a source of alpha particles)

- **Build** a radon detector and use it to detect radon emissions.
- **Observe** the tracks of alpha particles microscopically and count them.
- **Calculate** the activity of radon.
- **Evaluate** radon activity over a large area using class data and draw a map of its activity.

INTRODUCTION

Materials (for teacher use only)

- 10 mL, 6.25 M NaOH, per lab group
- 1000 mL beaker or 400 mL beakers, 2
- Bunsen burner or hot plate
- small test tube, 1 per lab group

Solution/Material Preparation

1. To prepare 1 L of 6.25 M NaOH, observe the required safety precautions. Slowly and with stirring, dissolve 250. g of NaOH in 750 mL of distilled water. Then, after the solution has cooled, add enough distilled water to make 1 L of solution. The NaOH solution will get hot as the pellets dissolve. To prevent boiling and splattering, add the pellets a few at a time with stirring.

2. An empty film canister can be used in place of the small plastic cup with lid.

3. CR-39 plastic is available from Alpha Trak, 141 Northridge Drive, Centralia, WA 98531, (206) 736-3884.

The element radon is the product of the radioactive decay of uranium. The $^{222}_{86}Rn$ nucleus is unstable and has a half-life of about 4 days. Radon decays by giving off alpha particles (helium nuclei) and beta particles (electrons) according to the equations below, with ^4_2He indicating an alpha particle and $^0_{-1}\beta$ representing a beta particle. Chemically, radon is a noble gas. Other noble gases can be inhaled without causing damage to the lungs because these gases are not radioactive. When radon is inhaled, however, it can rapidly decay into polonium, lead, and bismuth, all of which are solids that can lodge in body tissues and continue to decay. The lead isotope shown at the end of the chain of reactions, $^{210}_{82}Pb$, has a half-life of 22.6 years, and it will eventually undergo even more decay before creating the final stable product, $^{206}_{82}Pb$.

$$^{222}_{86}Rn$$
$$\searrow$$
$$^{218}_{84}Po + {}^4_2He$$
$$\searrow$$
$$^{214}_{82}Pb + {}^4_2He$$
$$\searrow$$
$$^{214}_{83}Bi + {}^0_{-1}\beta$$
$$\searrow$$
$$^{214}_{84}Po + {}^0_{-1}\beta$$
$$\searrow$$
$$^{210}_{82}Pb + {}^4_2He \rightarrow$$

In this experiment, you will measure the level of radon emissions in your community. First you will construct a simple detector using plastic (CR-39) that is sensitive to alpha particles. You will place the detector in your home or somewhere in your community for a 3 week period. Then, in the lab, the plastic will be etched with sodium hydroxide to make the tracks of the alpha particles visible. You will examine the tracks to determine the number of tracks per cm^2 per day and the activity of radon at the location. Finally, all class data will be pooled to make a map showing radon activity throughout your community.

SAFETY

  Wear safety goggles and a lab apron to protect for your eyes and clothing. You may remove your goggles only while you are using the microscope.

 Scissors and push pins are sharp; use with care.

MATERIALS

Etching Instructions

1. If students observe the etching process, they should wear safety goggles, a face shield, and a lab apron, and they should stand at least 5 ft from the caustic solution.

2. Pour enough 6.25 M NaOH into each test tube so that the plastic will be completely immersed when the paper clip is hooked over the test tube's edge.

- clear plastic ruler or stage micrometer
- CR-39 plastic
- etch clamp (the ring from a key chain)
- index card, 3 in. × 5 in.
- microscope
- paper clips
- push pin
- scissors
- small plastic cup with lid
- tape
- toilet paper or other tissue

Optional

- sources of radiation (Fiestaware, Coleman green-label lantern mantles, old glow-in-the-dark clock or watch faces, cloud-chamber needles)

PROCEDURE

3. Check the paper clip and etch clamp to be certain the students have attached them properly to the CR-39 plastic.

4. Hook the paper clip over the test tube lip so that the CR-39 plastic is completely immersed in the NaOH.

5. Add water to the beaker(s). Place the test tubes in the water.

6. Heat the water bath to boiling, and boil for about 30 min. The water may have to be replenished periodically.

7. Carefully remove the CR-39 from the hot-water bath, and rinse it thoroughly. Dry the plastic with a soft tissue.

Detector Construction

1. Cut a rectangle, 2 cm × 4 cm, from the index card.

2. Locate the side of the CR-39 plastic that has the felt-tip marker lines on it. Peel off the polyethylene film, and use the push pin to inscribe a number or other identification near the edge of the piece. With a short piece of transparent tape, form a loop with the sticky side out. Place the tape on the back of the piece of CR-39 plastic (the side that is still covered with polyethylene), and firmly attach it to the index-card rectangle.

3. With a permanent marker, write the ID number or other identification on the outside of the plastic cup. Place the paper rectangle, with the CR-39 plastic on top, into the cup.

4. Cut a hole in the center of the plastic-cup lid. Place a small piece of tissue over the top of the cup to serve as a dust filter. Then snap the lid onto the cup.

5. Place the completed detector in the location of your choice. (Check with your teacher first.) Record this information in Data Table 1. The detector must remain undisturbed at that location for at least 3 weeks.

6. At the end of the 3 weeks, return the entire detector to your teacher for the chemical-etching process and the counting of the radiation tracks.

Etching

7. Remove the CR-39 plastic from the plastic cup, detach it from the index card, and peel the polyethylene film from the back. Slip the ring of an etch clamp over the top of the CR-39 plastic, and hook it onto a large paper clip that has been reshaped to have "hooks" at each end, as shown in Figure A.

8. When you have completed this work, give the plastic to your teacher for the etching step. During this step, NaOH solution will be used to remove the outer layer of the plastic so that the tracks of the alpha particles become visible.

Counting the Tracks

9. Examine Figure B below. Notice the various shapes of the tracks left as alpha particles entered the CR-39 plastic. The circular tracks were formed by alpha particles that entered straight on, and the teardrop-shaped tracks were formed by alpha particles that entered at an angle.

10. Place your plastic sample under a microscope to view the tracks. Use a clear-plastic metric ruler or a stage micrometer to measure the diameter of the microscope's field of view. Make this measurement for low power (10×). Record this diameter in Data Table 1.

11. The tracks are on the top surface of the CR-39 plastic. Make certain that you focus on that surface and the tracks look like those in the illustration. These tracks were produced by placing the CR-39 plastic within a radium-coated clay urn. Your radon detector should not have nearly as many tracks. If there are too many tracks, switch to a high-power (40×) objective, and measure and record the diameter of the field of view. Place your piece of CR-39 plastic on a microscope slide. Count and record the number of tracks in 10 different fields. Record these numbers in Data Table 2.

FIGURE A

Alpha particle tracks

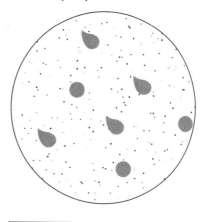

FIGURE B

Procedural Tips

• Thoroughly discuss all safety precautions outlined in this laboratory and in the safety section. Radioactivity is a poorly understood topic, so take this opportunity to discuss practical ways to reduce exposure.

• Discuss how the properties of radon as a gas cause problems in airtight buildings.

Data Table 1

Date detector was put in place	12/3/97
Time detector was put in place	3:45 PM
Location of detector (e.g., bedroom, kitchen, dining room, furnace room, under the stairwell, etc.)	kitchen
Floor level of location	1st
Date detector was removed	12/27/97
Time detector was removed	8:15 PM
Diameter of microscope field	Data will vary.

Data Table 2

Field	1	2	3	4	5	6	7	8	9	10
Number of tracks	3	4	4	5	9	7	4	3	8	6

CALCULATIONS

• Explain that the CR-39 plastic is covered with the polyethylene tape to protect it from radiation until it is ready for use. The NaOH bath strips away a thin layer of the plastic to expose the pathway of the alpha particles.

• Either demonstrate how to build the detector, or have one on hand for students to examine.

• You should not need to explain the usage of the microscope, but students will need help measuring and calculating the field of view. Also be certain that they focus the microscope on the proper level of the plastic.

• Discuss the calculations for Bq/L before the students make the calculations themselves. They often do not realize that *activity* = 2373 tracks/cm²/day was determined as a standard from 370 pCi/L of air (13.69 Bq/L). The standard was used to calculate unknown concentrations of radon.

1. **Organizing Data** In order to increase the accuracy of your data, you counted tracks in 10 different areas. Find the average number of tracks in a single area for your piece of CR-39 plastic.

$$\frac{3 + 4 + 4 + 5 + 9 + 7 + 4 + 3 + 8 + 6}{10 \text{ fields}} = 5.3 \text{ tracks/field on average}$$

2. **Organizing Data** You counted the number of tracks within the field of view of your microscope. In order to calculate the number of tracks per cm², you need to know the area of the field. Use the diameter of the field to calculate the area. (Hint: The field is circular; $d = 2r$; $A = \pi r^2$.)

$$A = \pi r^2 = (3.14)\left(\frac{0.17 \text{ cm}}{2}\right)^2 = 2.3 \times 10^{-2} \text{ cm}^2$$

3. **Organizing Data** Using the answers to Calculations items **1** and **2**, calculate the average number of tracks per cm².

$$\frac{5.3 \text{ tracks}}{0.023 \text{ cm}^2} = 230 \text{ tracks/cm}^2$$

4. Organizing Data In Calculations item **3** you calculated the number of tracks that accumulated per cm^2 over the total period the detector was in place. Divide this number by the number of days the detector was in place to calculate tracks/cm^2/day.

$$\frac{230 \text{ tracks/cm}^2}{24.19 \text{ days}} = 9.5 \text{ tracks/cm}^2\text{/day}$$

5. Organizing Data Several pieces of CR-39 plastic were sent to a facility that had a radon chamber in which the activity of the radon was known to be 13.69 Bq/L (becquerels per liter of air), as measured by a different technique. A becquerel is the name for the SI unit of activity for a radioactive substance and is equal to 1 decay/s. The exposed pieces of CR-39 plastic were etched, counted, and found to have an activity of 2370 tracks/cm^2/day. Calculate the radon activity measured by your detector in becquerels/liter. (Hint: Use the proportion 13.69 Bq/L : 2370 tracks/cm^2/day as a conversion factor to convert your data.)

$$9.5 \text{ tracks/cm}^2\text{/day} \times \frac{13.69 \text{ Bq/L}}{2373 \text{ tracks/cm}^2\text{/day}} = 5.5 \times 10^{-2} \text{ Bq/L}$$

6. Relating Ideas The Environmental Protection Agency and other U.S. government agencies use non-SI units of picocuries per liter (pCi/L) to measure radiation. One curie is 3.7×10^{10} Bq, and one picocurie is 10^{-12} curies. Convert your data into units of pCi/L.

$$5.5 \times 10^{-2} \text{ Bq/L} \times \frac{1 \text{ Ci}}{3.7 \times 10^{10} \text{ Bq}} \times \frac{10^{12} \text{ pCi}}{1 \text{ Ci}} = 1.5 \text{ pCi/L}$$

7. Organizing Data Combine your data with those of other teams in your class, and jointly construct a map of your region that shows the levels of radon activity in Bq/L and pCi/L at various locations throughout your community.

GENERAL CONCLUSIONS

1. Applying Ideas The half-life of $^{222}_{86}$Rn is 3.823 days. After what time will only one-fourth of a given amount of radon remain?

One-fourth will remain after 2 half-lives, 7.646 days.

2. **Designing Experiments** Factors other than geographic location can have an effect on radon emissions. For example, readings taken in basements are likely to be higher than those in attics. Design an experiment to test different aspects of the three highest and three lowest regions of radiation on the map to examine the influence of these other factors. If your teacher approves your suggestion, try it.

Students' suggestions will vary. Be sure that student plans meet

all necessary safety guidelines before allowing students to try

them.

3. **Designing Experiments** CR-39 plastic could be used for further investigations of naturally occurring radiation. Design an experiment to explore one of the following areas. If your teacher approves your plan, carry out the experiment.

- the range of alpha particles
- the radon activity in soil
- radioactivity of common items such as lantern mantles (Coleman green label), glow-in-the-dark clock or watch faces, and pieces of Fiestaware

Students' suggestions will vary. Be sure that student plans meet

all necessary safety guidelines before allowing students to try

them.

POWER CODERS

THE DAY OF THE GAMER

AMANDA VINK

ILLUSTRATED BY JOEL GENNARI

PowerKiDS press.

New York

Published in 2019 by The Rosen Publishing Group, Inc.
29 East 21st Street, New York, NY 10010

First Edition

Illustrator: Joel Gennari
Interior Layout: Tanya Dellaccio
Editorial Director: Greg Roza
Coding Consultant: Kris Everson

Cataloging-in-Publishing Data
Names: Vink, Amanda.
Title: Day of the gamer / Amanda Vink.
Description: New York : PowerKids Press, 2019. | Series: Power coders
Identifiers: ISBN 9781725301801 (pbk.) | ISBN 9781725301825 (library bound) | ISBN 9781725301818 (6pack)
Subjects: LCSH: Computer programmers–Juvenile fiction. | Coding theory – Juvenile fiction. | Graphic novels–Juvenile fiction.
Classification: LCC PZ7.1.V58 Da 2019 | DDC [F]–dc23

Manufactured in the United States of America

CPSIA Compliance Information: Batch CWPK19. For Further Information contact Rosen Publishing, New York, New York at 1-800-237-9932

CONTENTS

Gross Anatomy 4

Diego's Graphic Novel. 6

BRAINS!. 10

Think Like a Computer 18

Back on Track 24

Tricks and Treats!. 29

4

I'M HAVING TROUBLE REMEMBERING ALL THESE BODY PARTS.

HOW DO YOU MEMORIZE THEM?

GREAT QUESTION, NAYA.

THE ANSWER IS PRETTY BORING: IT'S JUST A MATTER OF STUDYING IT OVER AND OVER.

BIOLOGY

SOMETIMES I WISH STUDYING WAS MORE FUN.

WHAT ARE YOU GOING TO BE FOR HALLOWEEN, PETER?

DUNNO YET.

THE MOST IMPORTANT THING IS THAT I GET LOTS OF CANDY.

THE DAY OF THE GAMER!
COMPETITION
FOR BEST ORIGINAL GAME
WINNING TEAM
GOES TO THE ALL-STATE
CODING CONFERENCE

DID YOU SEE THE FLYER? I THINK WE SHOULD COMPETE.

SOUNDS FUN.

BUT WHAT KIND OF GAME?

7

I HAVE AN IDEA!

WE WANT TO MAKE A GAME FOR THE CODING COMPETITION.

DIEGO, CAN WE USE YOUR GRAPHIC NOVEL AS INSPIRATION?

YEAH, AWESOME!

ANOTHER PLUS...

MAYBE I'LL ACTUALLY BE ABLE TO MEMORIZE SOME BODY PARTS.

HOW DO YOU WANT TO PUT THIS TOGETHER?

WAIT!

NEED TO GET FUELED UP.

OK, READY.

WHY DON'T YOU MAKE A GAME WHERE YOU HAVE TO CHOOSE THE BODY PARTS FOR DR. LEGBONES?

YEAHHHHH.

I LIKE IT, BUT I THINK IT SHOULD ALSO BE EDUCATIONAL.

I AGREE.

YOU HAVE TO CHOOSE THE RIGHT BODY PARTS.

THERE SHOULD BE QUESTIONS ABOUT THE BODY PARTS.

IF YOU ANSWER CORRECTLY, THEN DR. LEGBONES CAN USE THAT BODY PART.

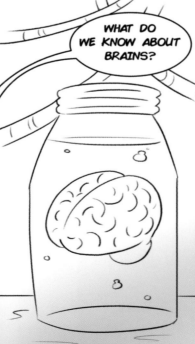

WHAT DO WE KNOW ABOUT BRAINS?

HUMAN ANATOMY

WE SHOULD BE ABLE TO FIND STUFF IN HERE.

9

THE BRAIN IS PART OF THE CENTRAL NERVOUS SYSTEM.

BRAINNNNSSSSSSSS!

AH!

HA HA HA HA HA

11

```
// This is where we create Dr. Legbone's requests! We're creating them as Javascript
//Objects to store and easily retrieve the questions/answers when we need them. Each
//Object (brainQuery) has two hints and one match message. 'keyName' is a custom
//identifier that we'll use in the evalItem() function to see if it matches the
//sprite selected by the user. We'll create the corresponding key names for the
//sprites in the next block of code.

var brainQuery = {
  hint:'Brains! I need brains!';
  missed:'Oops! I think my assistant needs some brains as well. Try again.';
  match:"Yes! Now that's using your head. The brain controls movement, writing,
reading, speech, and memory.";
  keyName:'brain';
  }
```

WOW, THAT'S IT?

FOR THE BRAIN!

BUT THERE'S A LOT MORE TO WRITE.

ONE FOR EVERY BODY PART AND TOOL.

YOU HAVE TO MAKE SURE EVERYTHING IS CORRECT TOO.

ONE MISTAKE AND THE WHOLE THING MIGHT NOT WORK.

WHILE YOU GUYS ARE CODING, DIEGO AND I WILL SCAN IMAGES FROM THE GRAPHIC NOVEL SO WE CAN USE THEM FOR OUR GAME.

PERFECT!

ROCK ON, POWER CODERS!

13

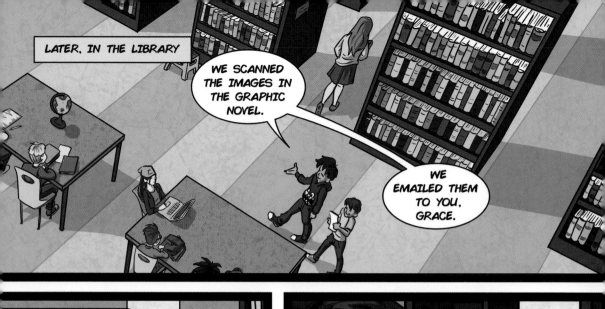

LATER, IN THE LIBRARY

WE SCANNED THE IMAGES IN THE GRAPHIC NOVEL.

WE EMAILED THEM TO YOU, GRACE.

PERFECT.

LET ME LOAD THEM NOW.

SO, WHAT DO WE DO NOW?

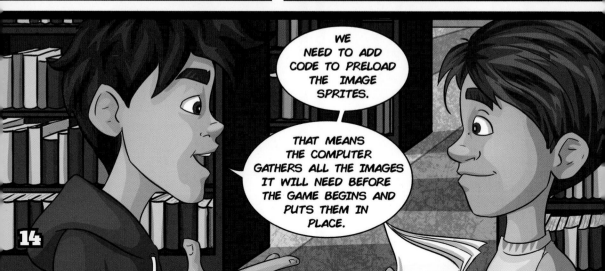

WE NEED TO ADD CODE TO PRELOAD THE IMAGE SPRITES.

THAT MEANS THE COMPUTER GATHERS ALL THE IMAGES IT WILL NEED BEFORE THE GAME BEGINS AND PUTS THEM IN PLACE.

14

THAT'S WHEN THE COMPUTER IS PRELOADING THE IMAGES.

THIS MAKES THE GAME WORK FASTER.

YOU'VE PROBABLY SEEN A "LOADING" SCREEN ON YOUR COMPUTER.

```
// preload sprite images!

preload {
    this.load.image('brain', 'assets/sprites/brain.png');
    this.load.image('skull', 'assets/sprites/skull.png');
    this.load.image('femur', 'assets/sprites/femur.png');
    this.load.image('intestines', 'assets/sprites/intestines.png');
    this.load.image('heart', 'assets/sprites/heart.png');
```

NOW THAT WE'VE LOADED THE IMAGE SPRITES, WE NEED TO MAKE THEM INTERACTIVE.

WAIT A SECOND...

I STILL DON'T UNDERSTAND WHAT AN IMAGE SPRITE IS.

IT'S A GRAPHIC IMAGE THAT CAN MOVE.

THE BRAIN IMAGE WE CREATED CAN BE CLICKED AND DRAGGED ALL AROUND THE PAGE.

COOL!

15

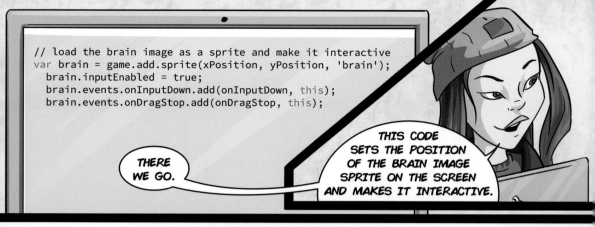

```
// load the brain image as a sprite and make it interactive
var brain = game.add.sprite(xPosition, yPosition, 'brain');
  brain.inputEnabled = true;
  brain.events.onInputDown.add(onInputDown, this);
  brain.events.onDragStop.add(onDragStop, this);
```

THERE WE GO.

THIS CODE SETS THE POSITION OF THE BRAIN IMAGE SPRITE ON THE SCREEN AND MAKES IT INTERACTIVE.

OK, I GET IT.

BUT WHAT'S WITH THE FIRST LINE?

THAT DOESN'T LOOK LIKE CODE.

THE FIRST LINE IS JUST A NOTE FROM THE CODER.

THE COMPUTER DOESN'T READ ANY LINE THAT STARTS WITH TWO FORWARD SLASHES.

IT TELLS OTHER PEOPLE WHAT THAT CODE DOES IN SIMPLE LANGUAGE.

OH WOW, THAT'S REALLY HELPFUL!

T. Legbones

BRRRRIIIIINNNNNGGGGG!

OH, I'VE GOT TO GO!

COOL.

WE CAN MEET TOMORROW TO TALK ABOUT OUR PROGRESS.

LET'S MEET AGAIN AFTER LUNCH.

MAKE SURE TO BRING YOUR GRAPHIC NOVEL!

I WILL!

SEE YA!

THE NEXT DAY

DO YOU KNOW WHAT YOU'RE GOING TO BE FOR HALLOWEEN, GRACE?

I'VE GOT AN IDEA.

I CAN'T WAIT TO SEE WHERE WE'RE AT WITH DR. LEGBONES.

HEY, DIEGO!

WAIT UNTIL YOU SEE WHAT WE CAME UP WITH LAST NIGHT.

18

Saint Malachy School Library

19

THIS IS LIKE A GIANT CODING PROBLEM.

HOW SO?

WELL, THE BRAIN IS SORT OF LIKE A GIANT COMPUTER.

YEAH!

ALAN TURING WAS A COMPUTER SCIENTIST IN THE 1940S.

HE THOUGHT A COMPUTER WOULD EVENTUALLY BE ABLE TO DO ALL THE COMPUTATIONS A HUMAN BRAIN CAN DO.

UH... OK?

A COMPUTER USES ELECTRICITY TO RECEIVE SIGNALS AND SORT THEM.

JUST LIKE A BRAIN USES NEURONS TO CREATE SYNAPSES.

WITH THE RIGHT INPUT, MAYBE YOUR BRAIN WILL REMEMBER THE OUTPUT.

WHAT ARE YOU SAYING?

22

DO YOU THINK MAYBE YOU DROPPED IT?

MAYBE MR. BARNES PICKED IT UP?

DIEGO, I WAS HOPING TO RUN INTO YOU TODAY.

MR. BARNES! DID I DROP MY GRAPHIC NOVEL?

YOU RAN AWAY SO FAST I COULDN'T CATCH YOU!

I'VE GOT IT RIGHT HERE.

YES!

OK, LET'S FINISH THIS CODE.

24

25

```
// an if/else conditional statement that decides whether or not the item
//selected and the item requested match

if (spriteKey === selectedKeyName) {

// It's a match! Show the 'match' message and load a new request from the Dr.

matchSuccess();
}
else {

// Wrong! Send sprite back to shelf and load the 'missed' message.

returnToOrigin(spriteSelected, originX, originY);
var wrongItemAlert = game.add.sound('wrong');
wrongItemAlert.play();
loadMissedMessage();
  }
}
```

28

HALLOWEEN

WHO GOT A 100?!

BIOLOGY

THIS GIRL.

NAYA, THAT'S AWESOME!

ALL BECAUSE I FINALLY MEMORIZED THE PARTS IN THE BODY, THANKS TO YOUR GRAPHIC NOVEL AND OUR GAME.

YOU GUYS!

WHAT'S UP, PETER?

THE RESPONSE IS HERE.

WHY ARE YOU FREAKING OUT?

IT'S UNOPENED.

BECAUSE... IT'S ALL-STATE!

LET'S OPEN IT.

WAIT!

CHOCOLATE MILK

OK.

WHAT?

31

32